Food and Packaging Interactions

ACS SYMPOSIUM SERIES **365**

Food and Packaging Interactions

Joseph H. Hotchkiss, EDITOR
Cornell University

Developed from a symposium sponsored
by the Division of Agricultural and Food Chemistry
at the 193rd Meeting
of the American Chemical Society,
Denver, Colorado,
April 5–10, 1987

American Chemical Society, Washington, DC 1988

Library of Congress Cataloging-in-Publication Data

Food and packaging interactions
    Joseph H. Hotchkiss, editor.

    p.   cm.—(ACS Symposium Series; 365).

    "Developed from a symposium sponsored by the
Division of Agricultural and Food Chemistry at the
193rd Meeting of the American Chemical Society,
Denver, Colorado, April 5–10, 1987."

    Bibliography: p.

    Includes index.

    ISBN 0–8412–1465–4
    1. Food—Packaging—Congresses.   2. Food—
Analysis—Congresses.

    I. Hotchkiss, Joseph H. II. American Chemical
Society. Division of Agricultural and Food Chemistry.
III. American Chemical Society. Meeting (193rd: 1987:
Denver, Colo.) IV. Series.

TP374.F64   1988
664´.092—dc19   88-1273
                                                   CIP

ACS Symposium Series

M. Joan Comstock, *Series Editor*

1988 ACS Books Advisory Board

Foreword

The ACS SYMPOSIUM SERIES was founded in 1974 to provide a medium for publishing symposia quickly in book form. The format of the Series parallels that of the continuing ADVANCES IN CHEMISTRY SERIES except that, in order to save time, the papers are not typeset but are reproduced as they are submitted by the authors in camera-ready form. Papers are reviewed under the supervision of the Editors with the assistance of the Series Advisory Board and are selected to maintain the integrity of the symposia; however, verbatim reproductions of previously published papers are not accepted. Both reviews and reports of research are acceptable, because symposia may embrace both types of presentation.

Contents

Preface

THE METHODS AND MATERIALS used to package food have changed more in the past 10 or 15 years than over the preceding 150 years since Appert invented the canning process. The technology has been driven mostly by marketplace needs and not by research and development.

The impact of these changes on the quality, safety, shelf life, and nutritional content of the packaged food has not been thoroughly researched. Only a few research groups have recognized that changes in food packaging technology can have an effect on the food itself. The desire for higher quality and safer food with a longer shelf life has led to increased interest in the interactions between foods and food packaging. The purpose of the symposium upon which this book is based was to bring together several of the leading research groups studying food and packaging interactions. Each group presented state-of-the-art discussions and many scientists pointed out areas in need of research. Others discussed concerns about new technologies or predicted changes.

The objective of this book is to not only summarize current work, but more importantly, to help set up an agenda for future research. It is my hope that the reports contained in this book will stimulate others to initiate research efforts. It is important that we understand the consequences, if any, of such fundamental and broad changes in the way we handle our food supply.

JOSEPH H. HOTCHKISS
Cornell University
Ithaca, NY 14853–7201

November 1987

Chapter 1

An Overview of Food and Food Packaging Interactions

Joseph H. Hotchkiss

Institute of Food Science, Food Science Department, Cornell University, Stocking Hall, Ithaca, NY 14853-7201

This paper is a brief overview of the symposium that was conducted by the Agricultural and Food Chemistry Division of the American Chemical Society at the Spring, 1987 meeting. Twenty-two papers were presented at the meeting on topics related to the interactions between foods and food packaging. Several papers dealt with specific topics while others were of a review nature. The objective of this introduction is to set the stage for the papers that follow in this volume.

There have been significant changes in both food processing and food packaging technologies over the last 5 to 10 years. These changes have included new ways to process foods, the use of new packaging materials, new combinations of standard materials, and new methods of manufacturing containers. While none of the basic materials (Table 1) used to package foods has escaped change, more change has occurred in the area of plastics than any other. For example, there has been a large increase in the use of plastic bottles for food packaging (Figure 1). Plastics which were once perceived as undesirable by food processors and consumers are now often seen as the best form of packaging available. Nearly all types of food packaging use plastics as part of their construction.

Change has also occurred in food processing over the last few years. The US Food and Drug Administration's 1981 approval of hydrogen peroxide to sterilize packages prior to filling was a watershed in food packaging development. The commercial success of this packaging process demonstrated that the US consumer was ready to accept packaging innovations if they provided a useful benefit. Soon after the success of aseptic packaging became apparent, development of new packages such as microwavable containers, retortable plastic cans, and selective/high barrier films were under development.

0097–6156/88/0365–0001$06.00/0
© 1988 American Chemical Society

TABLE I. Dollar value and market share
of materials used to manufacture packaging
(data from Rauch, 1986)[a]

Material	Year		
	1984	1985	1990[b]
Paperboard/pulp	18,842 (35)	19,340 (35)	24,580 (34)
Metal	14,638 (27)	15,052 (27)	18,600 (26)
Plastics	9,672 (18)	10,255 (18)	15,970 (22)
Paper	4,432 (8)	4,690 (8)	5,610 (8)
Glass	3,850 (7)	4,100 (7)	5,225 (7)
Wood	1,790 (3)	1,826 (3)	2,003 (3)
Textile	574 (1)	570 (1)	622 (1)
Total	53,798	55,833	72,610

a. millions of dollars (percent market share)
b. estimated

Innovations in food processing have placed new demands on
packaging and have accelerated the development of new packages.
For example, the desire to thermally process low acid foods in
rigid plastic cans has led to the development of multi-layered
plastic materials that maintain their barriers even after being
thermally processed in steam (1). Because these plastic
containers will most likely have flexible film closures instead of
rigid double seamed closures, heat sealing technology has also
become of critical importance.

These and other innovations are driven by economic and
marketplace forces. Food manufactures have found that packaging
can give them a competitive edge in the marketplace and consumers
have shown a willingness to pay more for package-product
combinations which offer greater convenience and/or higher
quality. The packaging industry has responded to food
manufactures by developing new packages that promise to offer
either reduced packaging costs or increased sales because of
added package convenience.   One of the most discussed examples
is the introduction of high barrier plastic bottles for oxygen
sensitive food products (2). Ketchup in this container was
reported to capture increased market share during its initial
introduction even though it sold for more money.

This introductory overview will discuss some examples of the
most recent changes in food processing and packaging technologies
and will point out how these changes have resulted in new problems
and opportunities for food manufactures. The individual papers
presented in this symposium deal in detail with the consequences
and/or needs created by these changing technologies. Research
into the interactions between foods and food packaging has lagged
well behind the development of new packages, but as this symposium

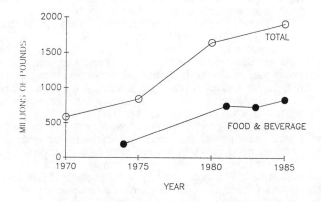

Figure 1. Growth in the production of plastic bottles for foods from 192 x 10^6 lbs in 1974 to 835 x 10^6 lbs in 1985 (Data from Rauch, 1986)

shows, is an area of growing research interest. Many foods that
were formerly packaged in nearly inert metal or glass containers
are now being packaged in less inert plastics and foods that were
once thought too sensitive for flexible films are now packaged in
high barrier flexible pouches. In fact, some foods such as "soft"
cookies or intermediate moisture foods might not be feasible
without such packaging. The interactions between foods and
packaging must now be considered when changing to less inert
materials, particularly plastics.

<u>Changes in Food Processing Technology</u>

Recently, consumers have shown a willingness to pay more for foods
that are perceived as fresher, higher quality, or of greater
value. Meeting this demand while maintaining adequate
distribution time has necessitated new technologies designed to
extend shelf life without sacrificing quality. Because of the
negative connotations of food additives, many shelf life extension
technologies are based on packaging.
    One of the most discussed methods to extend the shelf life of
refrigerated, perishable foods is modified atmosphere packaging
(MAP) (<u>3</u>). In this technology, the gases within a package are
altered to something other than air. This has two consequences.
First, changing the atmosphere changes the balance of
microorganisms. This can result in considerable
extension of shelf life for those products whose major mode of
deterioration is mediated through microorganisms. Secondly, the
respiration rate of fruits and vegetables can be reduced by a
change in atmosphere. This can greatly extend shelf life and may
result in the brand labeling of many products that are now
marketed as commodities.
    In order for MAP to be commercially successful, film
manufacturers will have to develop a wider range of selective
barrier films. For example, a film in which the permeation rate
of carbon dioxide was as low as that of oxygen would be useful in
maintaining internal atmospheres. Food manufacturers will have to
carefully determine what mixtures of gases successfully extend the
shelf life of their products
    The wide spread use of the home microwave oven has also meant
substantial changes in food packaging. Several major food
processors have converted their packages from metal to microwave
transparent packaging. Glass and many plastics are microwave
transparent and new packages made from each material have been
developed. Some packages incorporate a microwave absorbing
material in order to aid in the transfer of heat to a food. The
use of the microwave oven has also meant that higher temperature
plastics such as crystallized polyethylene terephthalate (CPET)
have had to be developed. These containers are not without
problems, however. Some containers can transfer undesirable odors
to foods during microwave cooking as discussed by Risch and
Reineccius in this volume.
    Food irradiation is controversial and its not clear if the
public is ready to accept irradiated foods. What is clear, is
that not enough is known about the effects of ionizing radiation
on some of the most useful packaging materials. As Thayer points

out in this volume, this is especially true for multilayered structures. The problems of polymer scission verses crosslinking, increases in the levels of potential migrants, and loss of strength or seal integrity have not been fully addressed. While there appear to be some materials available which will perform satisfactorily under irradiation, these materials may not have the desired barrier properties for long term storage of irradiated foods.

Aseptic packaging represents a true synergistic marriage of food processing and packaging technologies. The conventional way to produce a shelf stable food is to first package the food in a hermetically sealed container followed by batch sterilization (Figure 2). In aseptic packaging, the package and the food are sterilized separately, usually by different methods. The food undergoes some type of continuous heating in a heat exchanger while the package may undergo chemical, thermal, or radiation treatments. The sterilized food and package are brought together and aseptically filled and sealed (Figure 2). Neither the food process nor the package sterilization would be of use alone. In aseptic packaging, sterilization techniques that would not be suitable for foods could be used for the package; with hydrogen peroxide, for example. It also means that the package need not withstand the same sterilization process as the food. Cheaper, paper-based materials can be used when other materials could be prohibitively expensive.

The commercial success of the initial paperboard based aseptically packaged juices and related products has lead to the development of second generation aseptic packages. Many of these packages are made from multi-layered, thermoformed polymer based materials. These containers are usually heat sealed with a flexible lid material (4).

It is likely that there will be an even greater increase aseptic packaging as the technology to commercially sterilize foods containing particulate matter becomes available. This may mean new forms of packaging. Low acid foods in aseptic packages will present unique problems because the safety of these packages will depend on the reliability of the heat seals. More reliable and secure heat sealing polymers that still can be opened easily will need to be developed as will better inspection systems and test methods for seal strength and integrity.

The examples given above are only a few of the changes in food processing. More changes are on the horizon. Many of these innovations will involve packaging. The distinction between food processing and food packaging will become increasingly blurred as the technologies merge.

Changes in Packaging Technology

The food industry and the packaging industry are economically closely allied. Nearly 53% of the packaging industry's sales are to the food industry (5). In recent years this alliance has gone beyond economics to joint R&D efforts to develop improved food packages. The packaging industry is now investigating how specific packaging materials interact with foods and the food industry is becoming more involved in the direct development of

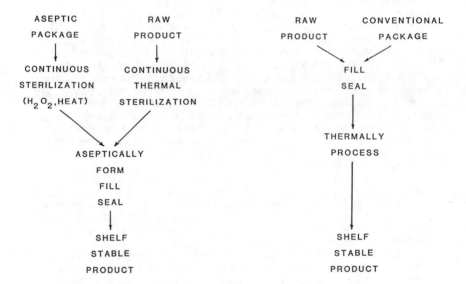

Figure 2. Comparison of conventional and aseptic processing
        systems for the production of shelf stable foods. In
        conventional processing the food is hermetically sealed
        in the package prior to processing. In aseptic
        processing the food and package are independently
        commercially sterilized prior to filling and sealing.

packaging. Some food companies have established their own pilot plant sized polymer processing facilities and nearly all major food companies have professionally staffed packaging departments.

This cooperation between the two industries has accelerated the development of new packaging technologies. For example, several container companies have developed the technology to co-extrude dissimilar plastics into sheets or parisons from which food containers can be formed. This has meant that high barrier polymers such as ethylenevinyl alcohol (EVOH) or poly(vinylidine dichloride) (PVDC) can be incorporated into a rigid tray or bottle, even though those materials would not be suitable by themselves (6). This technology has recently been used to package such oxygen sensitive foods as ketchup and mayonnaise. Modifications of the same technology are being used to manufacture rigid plastic "cans" that can withstand retorting at 121 C without losing their barrier properties. Several new food products have recently been introduced in these containers including soups, stews, and other entree items (7). The use of plastic containers for hot-filled and retorted foods can be expected to increase (8).

The food and packaging industries are also combining R&D efforts to make use of the dynamics of the interactions between food products and packages made from films. Two areas are emerging. First is the prediction of shelf life based on the barrier properties of the package and the rate and mode of deterioration of the food. By modeling these interactions, it is possible to predict shelf life and to optimize the package for the shelf life required by the product. This can result in reduced packaging costs. Several models for these interactions are being proposed and improved (See papers by Chao and Rizvi, and Taoukis and Labuza in this volume for a detailed discussion).

The packaging industry is also coming closer to being able to engineer desired permeability into films. This development may mean that respiring produce can be packaged in bags in which the internal atmosphere will quickly come to equilibrium. By selecting the proper package permeability, an internal atmosphere can be selected which will decrease the respiration rate of the produce. This will be similar to the controlled atmosphere storage of apples that has been practiced for several years.

Developments have not been restricted to the use of plastics, the glass and metal packaging industries have also developed new containers. The glass industry has taken advantage of the microwave transparency of glass and has worked with the food industry in developing new product-package combinations. The glass industry has also improved the way that glass containers are manufactured in order to reduce the cost of glass. The principal developments have been in making glass containers lighter in weight.

Metal can manufacture has substantially changed over the last few years. Metal cans are no longer manufactured by tin-coated, three-piece, side soldered techniques. A majority of cans produced in the US are now made without lead solder or tin. Most modern cans are either made from two pieces (a body and one end piece) or are made from three pieces with a welded side seam. As Good points out in this volume, this has necessitated the development of improved can interior coatings.

Food and Food Packaging Interactions

Changes in food packaging have meant that the ways foods interact
with packaging have likewise changed. Interactions between foods
and packaging can be classified into four types:

Migration or the Transfer of Components of the Package to the Food
During Storage or Preparation. Migration can have both quality
and toxicological significance. Very often, the components that
migrate from plastics are odor active and can adversely affect the
flavor of foods. This is especially a problem when foods are
heated in plastic containers, as in a microwave oven (See Risch
and Reiniccius in this volume). Migration may also result in the
transfer of potentially toxic substances to foods. There has been
considerable research in this area in recent years, yet all the
concerns have not been fully addressed (9). For example, there is
still concern about the transfer of vinyl chloride monomer (VCM)
to foods packaged in polyvinyl chloride (PVC). Migration of
potentially toxic components becomes a regulatory concern and at
least two papers (Breder, and Hollifield and Fazio) in this
symposium address the issues surrounding regulation of migrants.
The theoretical aspects of migration are addressed by Chang and
Smith in this volume.
     In recent years, migration has been used to transfer desired
additives to foods. At least one can manufacturer has developed a
system in which metal ions that will help stabilize the green
color of chlorophyll are incorporated into the can interior
coating. The desire to have antioxidants migrate from packaging
to foods during storage is also addressed in this volume (see
Harte et al).

Permeation of the Food Container to Fixed Gases and Water Vapor.
Unlike glass or metal containers, fixed gases and water vapor can
permeate packages made from plastics or thin foils. While all
plastics are permeable to some degree, permeation rates vary over
three orders of magnitude (10). Considerable work has been
undertaken in the area of predicting the effects of package
permeation on the shelf life of individual products. This work is
of considerable economic importance as the shift to plastic
packaging continues. In nearly all cases, higher barrier films
are much more expensive. The ideal situation is to package
products in materials which will protect foods only for the
maximum shelf life desired or found in the marketplace.
Protecting foods for longer periods than necessary is a form of
over-packaging (See Taoukis and Labuza in this volume for a
discussion).
     The objective of modeling shelf life work is to understand
how specific barriers influence the quality of individual foods.
There are at least three ways to approach this problem.   First
is to package the food in several containers and to determine the
shelf life of each under actual distribution conditions. While the
data from such a test is good, this is not a practical solution
because of the time necessary to conduct the test on foods that
may have a shelf life of several months. The second method is to
place the packages in storage under elevated temperature and

relative humidity conditions. This test is commonly conducted at 100 F and 90% relative humidity. This method decreases the time necessary for the test but problems occur when trying to relate the accelerated conditions to actual use conditions.

The third method is to combine the barrier properties of the package with the stability of the food in an appropriate mathematical model. This technique allows the shelf life to be predicted given any type of barrier and storage conditions. Several models have been proposed and several are reviewed by Chao and Rizvi in this volume.

Sorption and/or Permeation by Organic Vapors.   Continuous polymer films are permeable to organic vapors in a similar manner to fixed gases (10). Transfer of organic vapors across polymeric food packaging could have two adverse consequences. First, packaged foods that are exposed to undesirable volatile odors during storage or shipment might pickup the odor. The classical case occurs when diesel odors or the aroma of laundry soaps are absorbed by foods because of improper storage or shipping.

The second type of problem can occur when the desirable aroma compounds associated with a particular food are diminished by being sorbed into or permeated through the package. This latter problem has only recently became an area of research interest. Several papers in this volume deal directly with this area. While the data are not complete, a couple of generalizations can be made. First is that some plastics can sorb or transfer sufficient aroma compound to be detected by human senses. The second generalization is that just as different polymers vary greatly in their permeation rates for fixed gases, they also vary over orders of magnitude in their permeation rates for organic vapors. This means that foods that are sensitive to changes in their aromatic flavor can be protected by switching to higher aroma barrier films. Several papers in this symposium deal with this problem.

The Forth Interaction Between Foods and Food Packaging Results From the Transparency of Many Food Packages to Light.   Light, particularly in the shorter wave lengths, can catalyze adverse reactions such as oxidation in foods. This may lead to discoloration, loss of nutrients, or the development of off-odors.

Conclusions

The acceleration in the switch from nearly inert packaging to more interactive synthetic polymers has brought forth a new interest in the interaction between foods and food packaging. In the last few years research in this area has begun to gain momentum. The major problems in studying these interactions are the lack of standard methodology, agreed upon models, and data concerning the actual changes in food quality.

Literature Cited

1.   Haggin, J. Chem. Eng. News. 1984, 62(8), 21.
2.   Dembowski, R.J. Food and Drug Packaging 1983, 47(10), 17.
3.   Silliker, J.H.; Wolfe, S.K. Food Technol. 1980, 34(3), 59.
4.   Dilberakis, S. Food and Drug Packaging 1987, 51(5), 22.
5.   Rauch, J.A. The Rauch Guide to the Packaging Industry; Rauch
     Associates, Inc: Bridgewater, NJ, 1986; p 1.
6.   Anon. Packaging 1984, 29(12), 70.
7.   Anon. Packaging Digest 1987a, 24(3), 42.
8.   Anon. Food and Drug Packaging 1987b, 51(6), 45.
9.   Crosby, N.T. Food Packaging Materials: Aspects of Analysis
     and Migration of Contaminants; Applied Science Publishers
     Ltd: Essex, England, 1981; p 190.
10. Yasuda, H.; Stannett, V. In Polymer Handbook; J. Brandrup
     and E.H. Immergut, Eds; John Wiley & Sons: New York, New
     York, 1975; p 111-229.

RECEIVED September 24, 1987

Chapter 2

Transport of Apple Aromas
in Polymer Films

P. T. DeLassus [1], J. C. Tou[1], M. A. Babinec [1], D. C. Rulf[1],
B. K. Karp[2], and B. A. Howell [2]

[1]Dow Chemical U.S.A., Midland, MI 48674
[2]Department of Chemistry, Central Michigan University,
Mt. Pleasant, MI 48859

Permeation of apple aromas, especially trans-2-hexenal, in three
polymer films are described. A low density polyethylene film was
found to be a poor barrier. A vinylidene chloride copolymer film
and a hydrolyzed ethylene-vinyl acetate (EVOH) film were found to be
excellent barriers when dry. The EVOH was greatly plasticized by
humidity. Plasticizing effects due to higher permeant concentration
or due to co-permeant concentration were not large. The permeation
data were separated into the diffusivity and the solubility
coefficient.

Flavor management has become an important concern for food
packaging. The package is expected to do more than deliver safe,
wholesome food. The food must taste good too. While flavor
management has always been part of the food packaging equation,
general expectations of quality have been heightened, and high value
convenience foods must defend their price.

Flavor management contains several parts in the traditional
glass and metal containers. Flavor changes can result from
interactions with heat and light. Oxygen in the head space can be
important. Furthermore, the container surface can cause chemical
changes either as catalyst or co-reactant.

The new plastic containers have more flavor management
variables. In addition to the mechanisms important to glass and
metal containers, plastic containers have three more concerns.
First, molecules from the environment can permeate the package wall
and enter the food. The molecule may be as simple as oxygen or as
complex as the floral essence from the laundry products in the next
aisle of the supermarket. Second, molecules from the package might
migrate to the food. These molecules might come from the plastic
itself or from a coating or an adhesive. Third, flavor molecules
can leave the food by permeation into and through the plastic walls
of the package.

This paper will discuss some recent advances in this area. A
new experimental technique will be described with data for several
important packaging variables.

0097–6156/88/0365–0011$06.00/0

Framework for Experiments

Aromas. Apple aromas were chosen for these experiments for several reasons. Only a few compounds dominate the aroma of apple. These compounds are available from chemical supply houses.
These compounds have easily distinguishable mass spectrograms.
Finally, apple is being packaged in rigid plastic packaging for popular consumption. The specific compounds for this study were trans-2-hexenal, hexanal, and ethyl-2-methylbutyrate.

Films. Three films were included in this study. Low density polyethylene (LDPE) was included as a representative polyolefin. It is not considered to be a barrier polymer. It has permeabilities to selected aroma compounds slightly greater than the permeabilities of polypropylene and high density polyethylene (1). A vinylidene chloride (VDC) copolymer film was included as an example of a barrier that is useful in both dry and humid conditions. The film was made from Dow experimental resin XU32024.13 which is a new material designed for rigid applications. A hydrolyzed ethylene-vinylacetate (EVOH) copolymer film was included as an example of a barrier film that is humidity sensitive. The polymer was 44 mole % ethylene.
These films were tested as monolayers about 2.5×10^{-5}m (=1mil) thick. They would typically be used in multilayer structures; however, the individual permeabilities can be combined to predict the behavior of multilayers.

Instrument. A new instrument was designed and built for studying the transport of aromas in polymer films (2). The instrument is described schematically in Figure 1. The gas handling section of the instrument contains the plumbing, aroma containers, and the experimental film. This enclosure is insulated, and the temperature can be controlled, $\pm$ 1 °C, up to about 150°C.

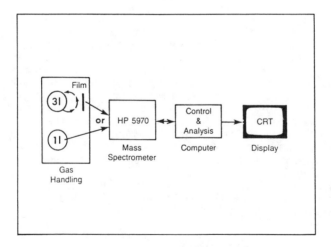

Figure 1. Schematic of Instrument

The detector was a Hewlett-Packard 5970 mass spectrometer. This instrument was used in two general ways. First, with each new compound, a total ion chromatogram was measured to identify the most populous and unique ion fragments. Second, for actual permeation experiments, the instrument was programmed to monitor selected ion fragments, typically three, for each of the aromas involved. When a single permeant was used, the three most populous ion fragments could be chosen. However, when mixed permeants were used, significant degeneracies were avoided. The response of the instrument was determined to be linear in the operating range and beyond, both at higher and lower concentrations.

In all experiments, the following sequence was used. First, a calibration experiment was completed by preparing a very dilute solution of the permeant in nitrogen in the one liter flask and introducing some of the gaseous solution directly to the mass spectrometer via a specially designed interface. A known concentration and a measured response yield a calibration constant.

Then a permeation experiment was completed by preparing a dilute solution of the permeant in nitrogen in the three liter flask and routing the gaseous solution past the purged, experimental film. The detector response was recorded as a function of time for subsequent analysis.

The computer was used to program and control the mass spectrometer and to collect and analyze the data. The computer was equipped with a CRT and a printer.

Other techniques have been used to measure flavor and aroma permation in polymer films. Gilbert (3) developed and still uses a quasi-static technique based upon a special film holder and gas chromatography. Giacin (4) uses a similiar technique and an improved isostatic method with a flame ionization detector. Murray (5) also uses a flame ionization detector in his quasi-isostatic method. Zobel (6) also describes a technique based upon a flame ionization detector. Caldecourt and Tou (7) developed isostatic techniques using either a photo-ionization detector or atomospheric pressure ionization/mass spectroscopy. Each of these techniques has a limitation either in detector sensitivity, requiring dry conditions, or allowing only one permeant at a time or high cost.

Variables. The experimental variables were chosen to make application to the real world. Temperature was chosen because packaged food encounters a variety of temperatures during the shelf life. While some packaged foods are refrigerated and some are stored at room temperature, many see unscheduled extremes in uncontrolled warehouses and unusual weather during transit, storage, and neglect. Usually elevated temperatures were used to enable timely completion of experiments and to obtain larger responses. Extrapolations will be discussed later.

Relative humidity was included as an experimental variable because it is a fact with food packaging. The package will be exposed to moisture from both the food and the environment. If a polymer is moisture sensitive, the relative humidity can make important contributions to performance.

The concentration of the permeant is important because a higher concentration of permeant can lead to a higher sorption by the package. A high concentration of permeant could lead to plasticization and high permeation for certain permeant/polymer combinations. Such effects have been reported (4,6).

A combination of permeants was included for several reasons. It is realistic. Food aromas are rarely built on a single compound. A combination tests to see if one compound either aids or hinders the permeation of another. It also provides another indirect test of plasticization.

Framework for analysis

For a thorough analysis of aroma transport in food packaging, the permeability (P) and its component parts - the solubility coefficient (S) and the diffusion coefficient or diffusivity (D) are needed. These three parameters are related as shown in Equation 1.

$$P = D \times S \tag{1}$$

The permeability is useful for describing the transport rate at steady state. The solubility coefficient is useful for describing the amount of aroma that will be absorbed by the package wall. The diffusion coefficient is useful for describing how quickly the permeant aroma molecules move in the film and how much time is required to reach steady state.

Equation 2 describes steady state permeation where ΔM_x is the

$$\frac{\Delta M_x}{\Delta t} = \frac{P \ A \ \Delta p_x}{L} \tag{2}$$

quantity of permeant x that goes through a film of area A and thickness L in a time interval Δt. The driving force for the permeation is given as the pressure difference of the permeant across the barrier, Δp_x. In an experiment, $\Delta M_x/\Delta t$ is measured at steady state while A and L are known and Δp_x is either measured or calculated separately.

The diffusion coefficient can be determined from the transient portion of a complete permeation experiment. Figure 2 shows how the transport rate or detector response varies with time during a complete experiment. At the beginning of an experiment, t=0, a clean film is exposed to the permeant on the upstream side.

Initially, the permeation rate is effectively zero. Then "break through" occurs, and the transport rate rises to steady state. The diffusion coefficient can be calculated in either of two ways. Equation 3 uses $t_{1/2}$, the time to reach a transport rate that is one-half of the steady state rate (8).

$$D = \frac{L^2}{7.2 \ t_{1/2}} \tag{3}$$

Equation 4 uses the slope in the transient part of the curve and the

$$D = \frac{0.176 \ L^2 \ (slope)}{Rss} \tag{4}$$

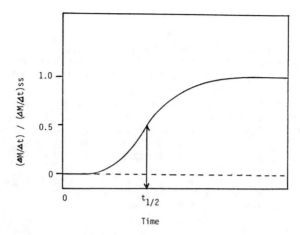

Figure 2. Relative Transport Rate as a Function of Time

steady state response Rss (9). After P has been determined with Equation 2 and D has been determined with Equation 3 (or 4), and then S can be determined with Equation 1.

Many sets of units are used to report the permeability in the literature. This paper will use SI units in the following way. For Equation 2 to be valid, the permeability must have dimensions of quantity of gas times thickness divided by area-time-pressure. If the kilogram is used to describe the quantity of gas and the pascal is used as the unit of pressure, then units of permeability are $kg \cdot m/m^2 \cdot s \cdot Pa$. This unit is very large and a cumbersome exponent often results. A more convenient unit, the Modified Zobel Unit, (1 MZU = 10^{-20} $kg \cdot m/m^2 \cdot s \cdot Pa$), was developed for flavor permeation. Zobel proposed a similar unit earlier (6). The units of the diffusion coefficient are m^2/s. The units of the solubility coefficient are $kg/m^3 Pa$.

For most simple cases, P, D, and S are simple functions of temperature as given in Equations 5, 6, and 7.

$$P\ (T) = P_0 \exp\ (-E_p/RT) \qquad (5)$$

$$D\ (T) = D_0 \exp\ (-E_D/RT) \qquad (6)$$

$$S\ (T) = S_0 \exp\ (-\Delta H_s/RT) \qquad (7)$$

where, P_0, D_0, and S_0 are constants, T is the absolute temperature, and R is the gas constant. E_p is the activation energy for permeation. E_D is the activation energy for diffusion, and ΔH_s is the heat of solution. If log P data are plotted on the vertical axis of a graph and T^{-1} is plotted on the horizontal axis, a straight line will result. The slope will be -0.43 E_p/R. The diffusion coefficient and solubility coefficient behave similarly. Equations 1, 5, 6, and 7 can be manipulated to yield equation 8 which relates the activation energies and the heat of solution.

$$E_p = E_D + \Delta H_s \qquad (8)$$

Equations 1-7 were developed for gas permeation through rubbery polymers. They must be used with caution with glassy polymers and/or organic vapors that interact strongly with the barrier.

Equations 5, 6, and 7 are valid above and below the glass transition temperature of the polymer, Tg, but not at Tg. Figure 3 shows a change in slope at Tg. Straight line extrapolations can not be made through Tg. For these studies of large permeant molecules in barrier films, P and D were very low at room temperature. This means that only a very low signal would result after many weeks. This was unacceptable.

Typically, experiments were done at several warm temperatures, and extrapolations were made to room temperature. This was no problem for the vinylidene chloride copolymer film (co-VDC film) since Tg is about 0°C. However, the EVOH film has a Tg at about 55°C. Extrapolation for this film was not allowed. The following procedure was used to estimate P and D below Tg. First, a simple extrapolation to room temperature of the high temperature data is made. This is shown as the lower dashed line in Figure 3. This yields a value for P or D which is almost certainly lower than the truth. Second, an extrapolation down to Tg followed by a horizontal translation is made. This is shown as the upper dashed line in Figure 3. Since E_p and E_D are positive quantities, this should yield a value for P or D which is almost certainly higher than the truth. The geometric average of the two extrapolations will be reported as an _arbitrary_ estimate of the transport parameters below Tg.

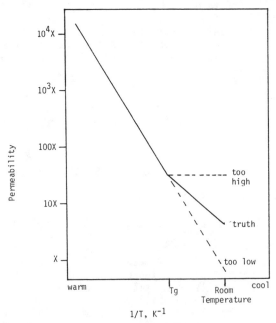

Figure 3. Temperature Dependence of Permeability, Extrapolating Below Tg

Experiments and Discussion

Typical experiment. Figure 4 shows the results of a typical experiment with the co-VDC film. The permeant was trans-2-hexenal. The experiment was done at 110°C. This warm temperature caused the experiment to end quickly. Three ions - the molecular ion m/z = 98 and two fragment ions m/z = 55 and 69 - were monitored by the mass spectrometer. The total ion intensity of these three ions was displayed. The first response is due to the calibration. Here 2.6 μl of a 0.5% solution of trans-2-hexenal in methylene chloride was injected into the 1 liter flask, diluted with nitrogen, and the resulting gaseous solution was introduced into the mass spectrometer via the interface.

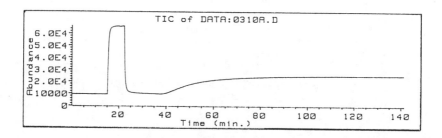

Figure 4. Permeation Experiment for Trans-2-hexenal in a Vinylidene Chloride Copolymer Film at 110 °C.

After the baseline was obtained again, at 34 minutes into the run, the permeation experiment began. Here 3.0 μl of trans-2-hexenal had been injected into the 3 liter flask and diluted with nitrogen before being circulated past the upstream side of the film. Breakthrough occurred at about 40 minutes, and steady state was reached at 80 to 100 min.

Figure 5 is a subset of the data from Figure 4. In Figure 5 only the data for the ion at m/z 55 is shown for calibration and permeation. Usually, the data for each ion are analyzed separately then averaged. Analyzing each ion separately is an internal check for consistency. Each ion should give the same result.

Concentration Effect. Some researchers have reported that the permeability of aromas in polymer films is dependent on the partial pressure of the permeant (4,6). One experiment was done here to test for this effect. Table I shows the result of this experiment. The permeant concentration of the upstream side of the polyethylene film is given in three ways. The first column is a recipe for the quantity of trans-2-hexenal injected into the 3 liter flask. The second and third columns express the concentration more traditionally as parts per million on a molar basis and the partial pressure of aroma in Pascals. Both were calculated with the ideal gas law.

In the case studied, the permeability was observed to be constant over this range of pressures. Experiments, in progress,

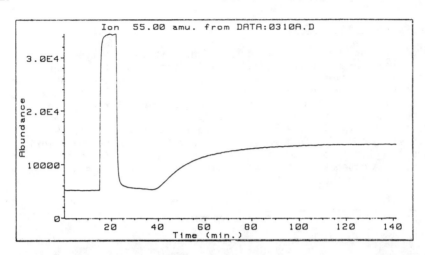

Figure 5.   Permeation Experiment for Ion 55 of Trans-2-hexenal
            in a Vinylidene Chloride Copolymer Film at 110°C.

Table I

PERMEABILITY OF TRANS-2-HEXENAL IN LOW DENSITY POLYETHYLENE AT 28°C

Permeant Concentration			Permeability
μl/3l	ppm	Pa	MZU
1	72	7.2	1.17×10^6
2	144	14.4	1.16×10^6
3	216	21.6	1.53×10^6
4	288	28.8	1.18×10^6

1 MZU = 10^{-20} kg•m/m^2•s•Pa

with trans-2-hexenal in a polar film are more likely to show a concentration effect.

Transport in a Vinylidene Chloride Copolymer Film. Figures 6 and 7 show the results of permeation experiments of trans-2-hexenal in the co-VDC film. The high temperatures were required to speed the experiments. Experiments at lower temperatures would have needed a much longer time. Each experiment used about 2 μl of permeant in the 3 liter flask. Table II contains the activation energies for these data in Equations 5, 6, and 7. Table III contains extrapolated permeabilities at selected temperatures.

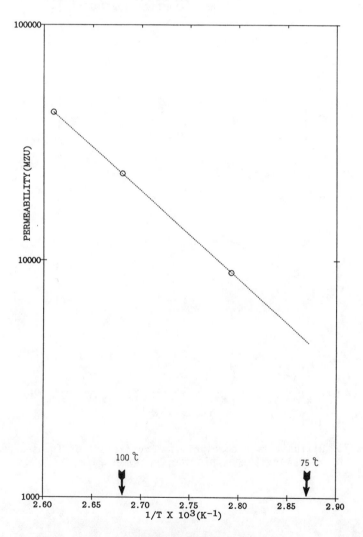

Figure 6.   Permeability of Trans-2-hexenal in a
Vinylidene Chloride Copolymer.

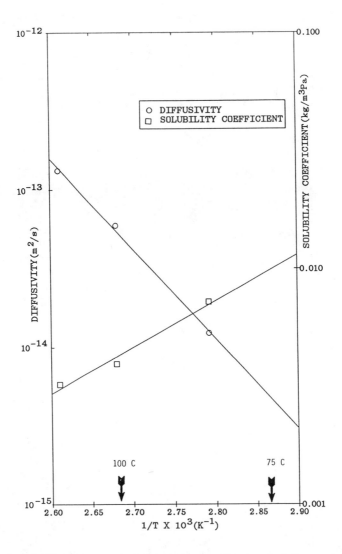

Figure 7. Diffusivity and Solubility Coefficient of Trans-2-hexenal
in a Vinylidene Chloride Copolymer Film.

Table II

MASS TRANSPORT OF TRANS-2-HEXENAL IN BARRIER FILMS

Vinylidene	E_p = 17.1 kcal/mole
Chloride	E_D = 26.1 kcal/mole
Copolymer	ΔH_s = -9.0 kcal/mole

EVOH (dry)	E_p = 17.0 kcal/mole
	E_D = 23.7 kcal/mole
	ΔH_s = -6.7 kcal/mole

Table III

EXTRAPOLATED[*] VALUES FOR TRANS-2-HEXENAL IN BARRIER FILMS

Film	Temperature °C	P MZU	D m^2/s	S kg/m^3 Pa
Vinylidene	75	4500	4.4×10^{-15}	0.01
Chloride	28	100	1.2×10^{-17}	0.08
Copolymer				
EVOH (dry)	75	2200	3.7×10^{-15}	0.006
	28	150	9.1×10^{-17}	0.02

*all values are the result of curve-fitting

Transport in a Dry EVOH Film. Figures 8 and 9 show the results of permeation experiments of trans-2-hexenal in the dry EVOH film. Again high temperatures were required to speed the experiments. Each experiment used about 2 μl of permeant in the 3 liter flask. Table II contains the activation energies for these data above Tg. Table III contains extrapolated permeabilities at selected temperatures.

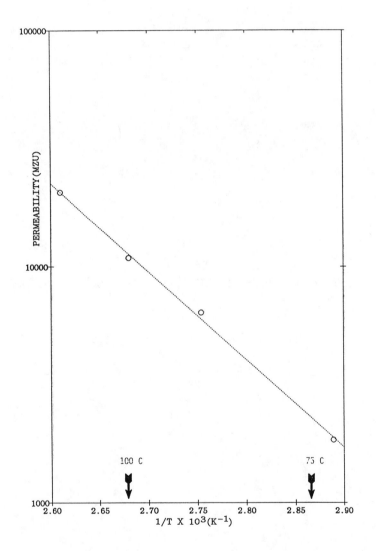

Figure 8. Permeability of Trans-2-hexenal in EVOH, Dry.

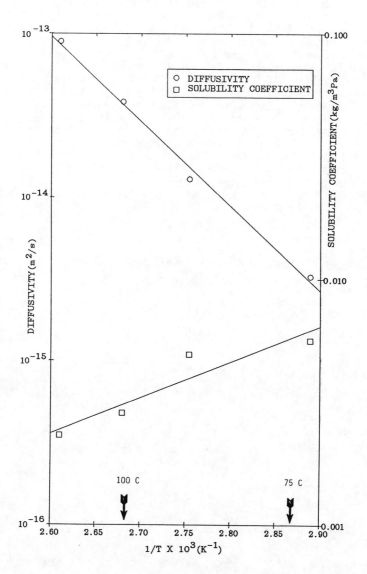

Figure 9.   Diffusivity and Solubility Coefficient of Trans-2-hexenal in EVOH, Dry.

<u>Humidity Effect</u>. For these experiments about 5 ml of a saturated solution of $ZnSO_4 \cdot 7 H_2O$ was put into the 3 liter flask. This held about 90% R.H. at the experimental temperatures (10). This humid nitrogen was circulated past the upstream side of the experimental film while dry nitrogen was routed past the downstream side of the film. After the film had equilibrated with the humidity for several hours, the permeation experiment was started.

The first humid experiment was done with trans-2-hexenal in the EVOH film at 75°C. The results are given in Table IV and are compared to the earlier dry data. Both the permeability and the diffusion coefficient increased by a factor of 45 while the solubility coefficient remained reasonably constant considering the experimental uncertainty when the film was humidified. This implies a simple plasticizing effect when the polymer is above Tg.

The second humid experiment was done with trans-2-hexenal in the EVOH film at 45°C. This experiment became possible because an increased diffusion coefficient shortened the experimental time and an increased permeability increased the signal strength. The results in Table IV are compared to an extrapolation of higher temperature data at dry conditions. At 45°C, which is below Tg, the

Table IV

EFFECT OF MOISTURE ON PERMEATION OF TRANS-2-HEXENAL
IN BARRIER FILMS

EVOH	Dry 0/0 RH	Moist 90/0 RH
Permeability, 75°C	2200 MZU	97000 MZU
Diffusivity, 75°C	1.6×10^{-14} m^2/s	7.2×10^{-13} m^2/s
Permeability, 45°C	330 MZU*	19000 MZU
Diffusivity, 45°C	2.6×10^{-16} m^2/s*	3.6×10^{-14} m^2/s

<u>Vinylidene Chloride Copolymer</u>

Permeability, 75°C	4500 MZU	4200 MZU
Diffusivity, 75°C	4.4×10^{-15} m^2/s	3.9×10^{-15} m^2/s

*extrapolated, all others are direct experimental data

permeability increased by a factor of 60 while the diffusion coefficient increased by a factor of 140. Hence the solubility coefficient decreased by a factor of 2. This implies a strong plasticizing effect from the moisture and possibly some competition for sorption sites when the polymer is below Tg. Within the framework of dual-mode sorption theory (11), the water molecules may compete for the Langmuir sorption sites better than the trans-2-hexenal can compete in this hydrophilic polymer. Whenever the permeant is excluded, the solubility coefficient decreases. This interpretation must be tempered by basic experimental uncertainty plus the unproven extrapolation technique. However, one fact is solid. The permeability of this flavor molecule in EVOH increases greatly under humid conditions.

The third humid experiment was done with trans-2-hexenal in the co-VDC film at 75°C. Table IV contains a comparison of the humid and dry experiments. Within experimental uncertainty, no difference was observed. This is consistent with other experience with films of Saran copolymers.(Saran is a trademark of The Dow Chemical Company). Many experiments in this laboratory have shown that the oxygen permeability in Saran copolymers is not affected by humidity.

Multiple Component Permeation. One experiment in four parts was done to test how three aromas might interact in permeation in polyethylene. Table V summarizes the results. First, a series of three tests was done. In each test a single aroma component was used at 2 μl in the 3 liter flask. The permeabilities were determined. Then, 2 μl of each of the three aroma components were put into the 3 liter flask, and a permeation test was completed. The permeabilities of each of the three aromas were determined. Table V shows that two of the aroma compounds increased in permeability in the multiple aroma test. One aroma compound decreased in permeability. The experiments proceeded too quickly to analyze the full transmission curve for anomalies. However, certainly there were no large interactions either by competition or plasticization. Other systems would need separate experiments. For this system, it appears that multiple component performance can be modeled with single component data.

Table V

MULTIPLE COMPONENT PERMEATION IN LOW DENSITY POLYETHYLENE

AT 28°C AND 144 PPM

Component	Permeability Alone MZU	Permeability Together MZU
Trans-2-hexenal	1.2×10^6	1.9×10^6
Hexanal	1.8×10^5	3.4×10^5
Ethyl-2-methylbutyrate	2.5×10^5	1.8×10^5

This experiment shows how a flavor imbalance could develop. The trans-2-hexenal permeates nearly ten times faster than the hexanal and the ethyl-2-methylbutyrate. Permeation during storage would cause the actual concentration of all three aromas to decrease; however, the relative concentration of trans-2-hexenal would decrease faster. The residual aroma in the food would be weak in one component. This might be unsatisfactory even if the total loss of aroma was within acceptable bounds.

SUMMARY

1. Table VI contains the best estimates of the permeability of trans-2-hexenal in polyethylene, EVOH, and a vinylidene chloride copolymer film. The parenthesis denote extrapolations of experimental data.

Table VI

PERMEABILITY OF TRANS-2-HEXENAL AT 28°C

	P(MZU)
LDPE	1.2×10^6
EVOH, Dry	(150)
EVOH, Moist	(>7000)
Vinylidene Chloride Copolymer Film	(100)

2. Low density polyethylene is not a good barrier to apple aromas.
3. Dry EVOH is a good barrier to trans-2-hexenal.
4. Moisture plasticizes EVOH greatly.
5. A vinylidene chloride copolymer film, wet or dry, is a good barrier to trans-2-hexenal.
6. The plasticizing effect of aromas is not clear.
7. The permeability, diffusivity, and solubility coefficient are strong functions of temperature. Small temperature changes can make large performance changes.

Literature Cited

1. DeLassus, P.T.; Hilker, B. L. Proc. Future-Pak '85 (Ryder), 1985, p 231.

2. Tou, J.C.; Rulf, D.C.; DeLassus, P.T., personal communication.

3. Gilbert, S.G.; Pegaz, D. Package January 1969, p 66.

4. Hernandez, R. J.; Giacin, J. R.; Baner, A. L. J. Plastic Film and Sheeting 1986, 2, 187.

5. Murray, L.J.; Dorschner, R.W. Package Engineering March 1983, p 76.

6. Zobel, M.G.R. Polymer Testing 1982, 3, 133.

7. Caldecourt, V.J.; Tou, J.C. TAPPI Proceeding, Polymers, Laminations and Coatings Conference 1985, p 441.

8. Ziegel, K. D.; Frensdorff, H. K.; Blair, D. E. J. Polym. Sci., Part A-2 1969, 7, 809.

9. Pasternak, R. A.; Schimscheimer, J. R.; Heller, J. J. Polym. Sci., Part A-2 (1970), 8, 467.

10. Carr D. S.; Harris, B. L. Ind. Eng. Chem. 1949, 41, 2014.

11. Vieth, W. R.; Sladek, K. J. J. Colloid Sci. 1965, 20, 1014.

RECEIVED December 12, 1987

Chapter 3

Sorption of *d*-Limonene by Sealant Films and Effect on Mechanical Properties

K. Hirose [1], B. R. Harte [1], J. R. Giacin [1], J. Miltz [2,3],
and C. Stine [2]

[1]School of Packaging, Michigan State University,
East Lansing, MI 48824–1223
[2]Food Science and Human Nutrition, Michigan State University,
East Lansing, MI 48824–1223

Absorption of d-limonene by sealant polymers
affected their: (1) modulus of elasticity; (2)
tensile strength; (3) ultimate elongation; (4)
seal strength; (5) impact resistance; and (6)
oxygen permeability. The degree of change caused
by absorption of flavor component varied with
each polymer.
     Strips of polyethylene and ionomer were
immersed in orange juice until equilibrium of d-
limonene absorption was established. The
mechanical properties and seal strength of the
control (non-immersed) and immersed samples were
determined at several concentration levels of
absorbant using a Universal Testing Instrument.
Oxygen permeability of the films increased due to
absorption of aroma constitutents.

Roughly 50 U.S. companies have introduced juice and juice drinks
in aseptic packages (1). Most aseptically filled juices are
packed into laminated carton packs, like the Brik Pak or Block Pak
(Combibloc).
     Danielsson (2) reported that Tetra Pak worldwide produced 35
billion packages using 580,000 metric tons of paperboard, 150,000
metric tons of polyethylene, and 30,000 metric tons of aluminum
foil in 1985. Sales of aseptic juice and juice drink now exceed
$250 million at retail (1) and are estimated to be $450 million by
the year 2000. Consumption is projected to increase to 3.9
billion gallons by the year 2000. Several juice companies are

[3]Current address: Pillsbury Company, Research and Development Laboratories,
Minneapolis, MN 55402; on leave from the Department of Food Engineering,
Technion–Israel Institute of Technology, Haifa, Israel

currently developing and test marketing large size aseptic packages (2 L to 4 L).

Further, many food processors are seeking to aseptically package not only new juice drinks but are also investigating the great potential for growth in other markets such as wines, milk products, puddings, gravies, sauces and soups. Thus, there will continue to be large quantities of aseptic cartons required for food and beverage packaging.

Shelf stability of orange juice is dependent upon the product and related critical factors, packaging being one. Most aseptically filled juices are packaged into laminated carton packs, like the Brik Pak or Combibloc, in which the usual sealant layer is polyethylene. Several authors (3-4-5) have reported that the shelf life of citrus juices in carton packs is shorter than in glass bottles. Other researchers (6-7) have pointed out that d-limonene, a major flavor component in citrus juices, is readily absorbed into polyethylene, thus reducing the sensory quality of citrus juices.

Therefore, characterization of the compatibility of polymer sealant films with aroma compounds would make it possible to select the more suitable packaging material. The objectives of this study were to investigate:

1. The absorption of d-limonene by carton packs containing aseptically packaged orange juice during storage at 24°C and 35°C.
2. The influence of d-limonene absorption on the mechanical and barrier properties of polyethylene and ionomer films as a function of d-limonene concentration.

MATERIALS AND METHODS
The Absorption of d-limonene by Carton Material in Contact With Aseptically Packed Orange Juice

Aseptically packed orange juice in laminated carton packages (polyethylene/kraft paper/aluminum foil/polyethylene -- Brik Style), were obtained from a local company. The packages contained 250 ml of juice each. A control was maintained by immediately transferring juice from the cartons to brown glass bottles at plant site. Quantitation of d-limonene was done according to the Scott and Veldhuis titration method (8).

To determine percent recovery, a known amount of d-limonene was weighed into a distillation flask and dissolved with 25 ml isopropanol. 25 ml of water was then added. The contents were then distilled (8) to determine the amount of d-limonene recovered.

To determine the amount of d-limonene in the carton material, the whole carton (1 x 0.5 cm pieces) was cut up after the juice was removed and the carton rinsed with distilled water. The carton pieces were then placed into a distillation flask. 70 ml isopropanol was added so that the sample was completely immersed. The sample/isopropanol mixture was allowed to sit for approximately 24 hr. Several extraction times were used to determine recoveries of d-limonene from carton stock. 70 ml of water was then added and the sample distilled as before.

Effect of d-limonene Absorption on Mechanical Properties of Polymer Films

The orange juice used in this study was 100% pure (from concentrate) obtained from a local company. The antioxidants, Sustane W and Sustane 20 (UOP Inc., 0.02% w/w total), and the antibacterial agent sodium azide (Sigma Chemical Co.) (0.02%, w/w) were added to the juice in order to prevent oxidative and microbial changes during storage.

To determine the effect of antioxidant and antimicrobial agents on juice stability, pH (Orion Research Co. Analog pH Meter, Model 301), juice color (Hunter D25 Color Difference Meter) and microbial total counts were monitored as indicators of juice quality.

Sealant Film Samples and Conditions

Low density polyethylene film (LDPE) (5.1×10^{-2} mm thick) was obtained from Dow Chemical Co., while Surlyn S-1601 (sodium type; 5.1×10^{-2} mm thick) and Surlyn S-1652 (zinc type; 7.6×10^{-2} mm thick) were obtained from DuPont. The samples were cut into strips, 2.54 cm x 12.7 cm, and immersed into juice (7 samples per 250 ml bottle) which had been filled into amber glass bottles and closed with a screw cap.

The ratio of film area to volume for a 250 ml pack (area/volume) was 0.9. Film strips were also heat sealed together using an impulse heat sealer and immersed into the juice (7 samples per bottle).

Stress-Strain Properties

Stress-strain properties were determined as a function of absorbant concentration using a Universal Testing Instrument (Instron Corporation, Canton, MA). The procedure used was adopted from ASTM Standards D882-83 (1984). Ten specimens were tested to obtain an average value. The amount of d-limonene absorbed was determined according to the Scott and Veldhuis (8) procedure.

Influence of d-limonene Absorption on Impact Resistance of Polymer Films

Sample specimens of the same LDPE (5.1×10^{-2} mm) and Surlyn S-1652 (7.6×10^{-2} mm) were immersed into orange juice until equilibrium (d-limonene) was established. The impact resistance of the control (non-immersed) and the sample films (immersed for 18 days) was measured using the free-fall dart method (ASTM Standard D 1709-85, 1986). Test Method A was used with a drop height of 0.66 m for LDPE, and 1.52 m for ionomer.

Influence of d-limonene Absorption on Barrier Properties of Polymer Films

Sample specimens of the same LDPE and Surlyn S-1652 were immersed into orange juice until equilibrium (d-limonene) was established. The oxygen permeability of the control (non-immersed) and immersed samples were measured using an Oxtran 100 Oxygen Permeability Tester (Modern Controls, Rochester, MN). Permeability measurements were performed at 100% RH and 23°C.

RESULTS AND DISCUSSION
Recovery of d-limonene Using standard d-limonene solutions (Sigma Chemical Co. Ltd.), percent recovery of d-limonene was found to range from 98.0 to 100.0. Recovery of d-limonene from carton stock ranged from 95.0 - 99.5. A carton extraction time of 24 hr proved to be adequate.

Distribution of d-limonene Between Juice and Carton Stock
Distribution of d-limonene between juice and carton stock in aseptically packed orange juice stored at 24°C, 49% RH and 35°C, 29% RH is shown in Table 1. The carton originally contained 2.5 mg d-limonene/package. The carton material had already absorbed some d-limonene when the samples were put in storage (Day 0) because product was obtained 1 day after packing.

A rapid loss of d-limonene in the juice, from 25 mg to 19.9 mg at 24°C and to 17.8 mg at 35°C, was observed within 3 days storage apparently due to absorption by the polyethylene contacting surface. After 3 days, loss of d-limonene from the juice proceeded at a lower rate. These results were similar to those found by other authors (6-7-9).

Rapid absorption of d-limonene into the polyethylene occurred during the beginning of storage. The rate of absorption then decreased. After 12 days of storage, saturation was reached. At equilibrium, the amount at 24°C and 35°C was about 11 mg/package (44% of the initial content) and 9.5 mg/package (38%), respectively (Table 1). The distribution ratio of d-limonene at equilibrium

$(\dfrac{\text{equilibrium content of d-limonene in carton stock}}{\text{equlibrium content of d-limonene in juice}})$ was about 0.65

24°C and 0.61 at 35°C, respectively. This suggests that the amount of d-limonene absorbed by the polyethylene and the rate of absorption were independent of storage temperature, within the temperature range studied.

Effect of d-limonene Absorption on Mechanical Properties of Sealant Films
Due to the addition of antioxidants and antimicrobial agents, no bacterial growth (< 50 counts/250 ml) and essentially no change in any of the color parameters, as measured by the Hunter Color Difference Meter, were detected. During 25 days storage, the pH decreased from 3.62 to 3.55. Therefore, both agents were added to the orange juice prior to conducting all studies with sealant films.

From work described previously, it was found that d-limonene was rapidly sorbed by the film contact layer. Therefore, to understand the influence of d-limonene absorption on mechanical properties of polymer films, strips of the packaging material were immersed in orange juice. The influence of absorption on modulus of elasticity, tensile strength, % elongation and seal strength were determined. The data were statistically evaluated by one way analysis of variance.

Absorption of d-limonene by the films during storage at 24°C, 49% RH is shown in Table 2. Within 3 days storage, all of the polymers films had rapidly absorbed d-limonene. After 3 days, the

Table 1

Distribution of d-limonene Between Juice and Package Material in Aseptically Packed Orange Juice During Storage at 24°C and 35°C*

	0	3		6		11		18		25	
Sample	Initial	24°C	35°C	24°C	35°C	24°C	35°C	24°C	35°C	24°C	35°C
Juice in glass bottle (mg/250 ml)	25.0	24.7	23.4	24.7	23.0	24.0	22.6	23.8	21.1	23.7	20.3
Juice in carton package (mg/250 ml)	25.0	19.9	17.8	18.6	16.9	17.2	16.5	16.9	14.0	14.0	12.1
Carton Package (mg/package)	2.5	7.5	7.6	9.9	8.6	10.7	9.2	11.6	10.1	11.6	10.6

Storage Time (days)

*Average of two determinations.

rate of absorption in LDPE and Surlyn (sodium type) decreased, while Surlyn (zinc type) had reached saturation. 12 and 18 days were required to reach saturation for Surlyn (sodium type) and LDPE respectively. The amount of d-limonene absorbed at equilibrium was 6.4 mg/100 cm^2 for Surlyn (sodium type), 5.3 mg/100 cm^2 for LDPE, and 3.3 mg/100 cm^2 for Surlyn (zinc type). The carboxy and zinc groups in the Surlyn probably alter the lipophilic character of the polymer, but do not prevent absorption of flavor components.

Table 2
Distribution of d-limonene Between Orange Juice and Sealant Films During Storage at 24°C, 49% RH

Content of d-limonene (mg)

Storage Time (days)	Juice	LDPE	Juice	S-1601	Juice	S-1652
0	46.3	0	46.3	0	35.5	0
3	36.1	6.8	29.3	10.8	26.6	7.0
6	33.9	8.3	26.6	12.0	23.4	7.3
12	28.5	10.2	23.5	14.7	23.2	7.4
18	26.6	11.8	22.5	14.7	22.4	7.5
27	26.5	11.9	22.3	14.5	21.6	7.5

The values are represented as mg/250 ml in juice and mg/225.8 cm^2 in films.
The results are the means of triplicates.
S-1601: Surlyn sodium type
S-1652: Surlyn zinc type

Modulus of Elasticity
The modulus of elasticity of the films, as a function of storage time, is shown in Table 3. Relative percent of modulus of elasticity is shown in Figure 1.

Results of statistical analysis showed that d-limonene absorption significantly affected the modulus of elasticity of LDPE and Surlyn (sodium type) (a = 1%). After 3 days storage, both materials had significantly lower values (a = 3%) in comparison with initial values. Further decreases were noted with increasing absorption of d-limonene in LDPE and Surlyn (sodium type). Retention of the modulus of elasticity by LDPE was greater than that for Surlyn (sodium type). The absorption of d-limonene decreased the stiffness of these two films. For the Surlyn film (zinc type), modulus of elasticity in both MD and CD was not affected significantly due to absorption of d-limonene.

Tensile Strength
The mean values for LDPE (MD) decreased due to absorption (a = 1%), although the degree of influence was less than that for Surlyn (zinc type) (Table 4 and Figure 2). There

Table 3

Modulus of Elasticity (1 x 10^7 Pa) of Films Immersed in Orange Juice During Storage at 24°C, 49% RH*

| | Storage Time (days) | | | | | | | | | | | |
| | 0 | | 3 | | 6 | | 12 | | 18 | | 27 | |
	$\bar{x}$	S.D.	$\bar{x}$	S.D.	$\bar{x}$	S.D.	$\bar{x}$	S.D.	$\bar{x}$	S.D.	$\bar{x}$	S.D.
LDPE (MD)[a]	6.76	0.20	6.40	0.15	6.03	0.19	6.03	0.21	6.09	0.17	6.06	0.10
LDPE (CD)[b]	7.48	0.61	6.53	0.22	6.38	0.14	6.42	0.15	6.34	0.14	6.32	0.15
S-1601[c] (MD)	5.87	0.12	4.12	0.50	40.4	0.51	3.93	0.24	3.57	0.55	3.48	0.34
S-1601 (CD)	5.79	0.12	3.92	0.45	3.86	0.51	4.01	0.45	4.29	0.39	3.85	0.43
S-1652[d] (MD)	6.94	0.15	6.94	0.15	6.94	0.15	6.89	0.12	6.83	0.20	6.89	0.12
S-1652 (CD)	7.22	0.38	6.85	0.26	6.89	0.24	6.86	0.15	6.89	0.31	6.89	0.24

*The results are means of 10 determinations.

[a]MD = Machine Direction

[b]CD = Cross Direction

[c]S-1601 = Surlyn – sodium type

[d]S-1652 = Surlyn – zinc type

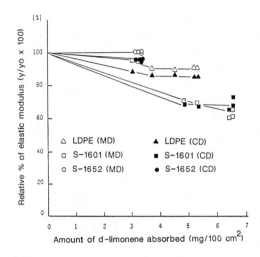

Figure 1. Relationship between d-limonene content and modulus of elasticity for test films.

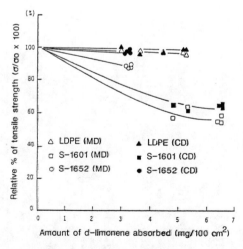

Figure 2. Relationship between d-limonene content and tensile strength for the test films.

Table 4

Tensile Strength (1 x 10^7 Pa) of Films Immersed in Orange Juice During Storage at 24°C, 49% RH*

| | Storage Time (days) | | | | | | | | | | | |
| | 0 | | 3 | | 6 | | 12 | | 18 | | 27 | |
	$\overline{x}$	S.D.	$\overline{x}$	S.D.	$\overline{x}$	S.D.	$\overline{x}$	S.D.	$\overline{x}$	S.D.	$\overline{x}$	S.D.
LDPE (MD)[a]	1.92	0.04	1.88	0.05	1.81	0.06	1.82	0.08	1.82	0.07	1.78	0.09
LDPE (CD)[b]	1.40	0.04	1.39	0.03	1.37	0.05	1.35	0.04	1.37	0.05	1.37	0.03
S-1601[c] (MD)	2.88	0.28	1.62	0.22	1.78	0.21	1.65	0.20	1.50	0.23	1.53	0.23
S-1601 (CD)	2.88	0.21	1.82	0.12	1.74	0.17	1.81	0.21	1.85	0.26	1.82	0.23
S-1652[d] (MD)	2.43	0.15	2.15	0.14	2.12	0.13	2.16	0.12	2.14	0.13	2.11	0.14
S-1652 (CD)	2.23	0.13	2.14	0.15	2.17	0.19	2.16	0.23	2.17	0.19	2.19	0.14

*The results are means of 10 determinations.

[a]MD = Machine Direction

[b]CD = Cross Direction

[c]S-1601 = Surlyn - sodium type

[d]S-1652 = Surlyn - zinc type

Table 5

% Elongation at Break of Films Immersed in Orange Juice During Storage at 24°C, 49% RH*

	Storage Time (days)											
	0		3		6		12		18		27	
	x̄	S.D.	x̄	S.D.	x̄	S.D.	x̄	S.D.	x̄	S.D.	x̄	S.D.
LDPE (MD)[a]	267 (100)	30.4	346 (129.6)	28.1	322 (120.6)	23.6	26.2 (117.6)	330	15.8 (123.6)	325	23.7 (121.7)	0.10
LDPE (CD)[b]	600 (100)	37.9	522 (87.0)	130	548 (91.3)	1304	493 (82.2)	119	512 (85.3)	111	430 (71.7)	147
S-1601[c] (MD)	451 (100)	35.1	429 (95.1)	35.1	446 (98.9)	47.4	421 (93.3)	37.4	413 (91.6)	45.4	436 (96.7)	49.6
S-1601 (CD)	449 (100)	30.4	426 (94.1)	30.4	436 (97.1)	24.0	433 (96.4)	34.5	4349 (96.7)	32.9	439 (97.8)	42.6
S-1652[d] (MD)	530 (100)	36.6	526 (99.2)	28.1	527 (99.4)	23.4	519 (97.9)	14.1	520 (98.1)	20.3	522 (98.4)	19.5
S-1652 (CD)	489 (100)	24.8	516 (105.5)	33.2	520 (106.3)	24.0	521 (106.5)	30.4	524 (107.2)	16.5	523 (107.0)	18.0

*The results are means of 10 determinations.
[a]MD = Machine Direction
[b]CD = Cross Direction
[c]S-1601 = Surlyn – sodium type
[d]S-1652 = Surlyn – zinc type

Table 6

Seal Strength (N/m) of Films Immersed in Orange Juice During Storage at 24°C, 49% RH*

	$\overline{x}$ 0	S.D.	$\overline{x}$ 3	S.D.	$\overline{x}$ 6	S.D.	$\overline{x}$ 12	S.D.	$\overline{x}$ 18	S.D.	$\overline{x}$ 27	S.D.
LDPE (MD)[a]	17.00	0.93	17.04	1.20	14.82	2.67	14.95	1.02	14.73	2.27	13.97	1.69
S-1601[b] (MD)	16.38	0.49	12.95	0.85	12.19	0.85	11.57	1.02	12.68	0.58	12.42	0.67
S-1652[c] (MD)	22.16	1.42	21.09	0.85	21.27	0.71	21.45	0.67	21.54	0.67	21.32	0.71

*The results are means of 10 determinations.

[a]MD = Machine Direction

[b]S-1601 = Surlyn - sodium type

[c]S-1652 = Surlyn - zinc type

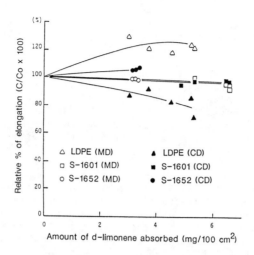

Figure 3. Relationship between d-limonene content and percent elongation at break for the test films.

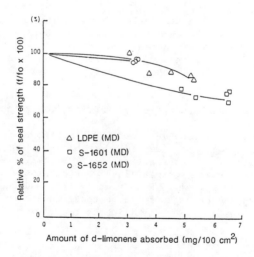

Figure 4. Relationship between d-limonene content and seal strength for the test films.

were no significant differences in CD of LDPE and Surlyn (zinc type). However, the tensile strength of Surlyn (sodium type) decreased to 53% and 63% of the original value which was similar to that found for the modulus of elasticity (59% - 66%).

% Elongation
The results (Table 5, Figure 3) show that percent elongation of LDPE (MD) at break increased due to absorption (a = 1%). The absorbant probably acted as a plasticizer to allow the chains to side post one another. Elongation of LDPE (CD) tended to decrease though the difference was not significant. The absorption of d-limonene by the surlin films did not affect their elongation.

Seal Strength
Change in seal strength due to d-limonene absorption for each film is shown in Table 6, Figure 4. The seal strength of Surlyn (sodium type) decreased due to absorption. A reduction of 24% in the seal strength was found after maximum absorption occurred (a = 1%), which was lower than the value found for LDPE. For Surlyn (zinc type), no significant decrease in seal strength was observed due to absorption.

Influence of d-limonene Absorption on Impact Resistance of Polymer Films
The impact failure for LDPE and Surlyn (zinc type) (non-immersed) was 69.0 g and 217.5 g, respectively. After d-limonene was absorbed by the films, the impact resistance of Surlyn (zinc type) increased to 244.5 g (12.4% increase). The failures were different, upon contact with the LDPE film, the dart caused a slit to tear the film, whereas in the Surlyn film it made a circular hole.

Influence of d-limonene Absorption of Barrier Properties of Polymer Films
The oxygen permeability constants for LDPE, Surlyn (sodium type) and Surlyn (zinc type) (non-immersed) were 88.9, 111.8, and 116.8 (cc $\cdot$ mm/ m^2 $\cdot$ day $\cdot$ atm), respectively. After d-limonene was absorbed by the films, the permeability constants were 406.4, 279.4 and 185.4 respectively. Mohney et al (10) and Baner (11) found that absorption of organic flavor constituents by test films increased the permeability of the films to the flavor constitutents.

REFERENCES

1. Sacharow, S. Prepared Foods, 1986. 155(1), 31-32.
2. Danielsson, L. Boxboard Containers, 1986, 93(6):25-26.
3. Gherardi, D.S. Proceedings of the International Congress of Fruit Juice Producers. 1982. Munich, Germany.
4. Mannheim, C.H. and Havkin, M. J. Food Proc. and Preserv., 1981, 5:1.
5. Granzer, R. Verpackungs-Rundschau, 1982, 33(9):35-40.
6. Marshall, M.R., Adams, J.P. and Williams, J.W. Proceedings Aseptipak '85, 1985, Princeton, NJ.

7. Mannheim, C.H., Miltz, J., and Letzter, A. J. Food Science. 52:741-746.
8. Scott, W.C. and Veldhuis, M.K. J.AOAC, 1966, 49(3):628-633.
9. Durr, P. and Schobinger, U. Lebensmittel Verpackung, 1981, 20:91-93.
10. Mohney, M.S., Hernandez, R.J., Giacin, R.J., Harte, B.R. 13th Annual IAPRI Symposium. 1986, Oslo, Norway.
11. Baner, A.L. 13th Annual IAPRI Symposium, 1986, Oslo, Norway.

RECEIVED July 1, 1987

Chapter 4

Permeation of High-Barrier Films by Ethyl Esters

Effect of Permeant Molecular Weight, Relative Humidity, and Concentration

J. Landois–Garza and Joseph H. Hotchkiss

Institute of Food Science, Food Science Department, Cornell University, Stocking Hall, Ithaca, NY 14853-7201

This paper reports the effects of changes in permeant molecular weight, relative humidity (RH) and permeant concentration on the permeability of alkyl esters through polyvinyl alcohol (PVOH) films, as measured by a differential permeation method. As the permeant molecular weight increased, the solubility coefficient (S) increased at a higher rate than the diffusion coefficient (D) decreased, resulting in an increased permeability coefficient (P). As the RH increased, S decreased, while D remained constant at moderate RH's and decreased at high RH's, resulting in a lower P. As the permeant concentration increased, D decreased while S appeared linear at low concentrations and increased non-linearly at higher concentrations, resulting in a non-linear increase of P.

The rapid growth in the use of barrier plastics for packaging food products that were previously packaged in glass or metal containers has caused an increased interest in the interaction between plastics and flavors and aromas. Unlike glass or metal, plastics can interact with foods and alter their original flavor and aroma. Flavor deterioration may occur by the ingress of foreign odors, the migration of plastic components into the food, or by the permeation and/or absorption of desirable aromas into the container walls.

Most studies on the permeability of plastic films to organic compounds have been performed using saturated vapors. The results obtained with this method may not be directly applicable to the permeation of the flavor and aroma components in foods, since they are present at very low concentrations and vapor pressures. The object of this work was to study the effect of permeant molecular weight, relative humidity, and permeant concentration on the permeability of organic vapors through a high barrier polymer.

0097–6156/88/0365–0042$06.00/0

THEORY

Permeability (P) depends on both the solubility and the rate at which permeant molecules diffuse through the polymer, and is a product of diffusion (D) and solubility (S). Hence, if any of two coefficients are known, the third may be obtained by:

$$P = D\,S \tag{1}$$

A steady state permeation exists when equal amounts of permeant enter and leave the polymer sheet. In the early stages of the process, more permeant dissolves into the membrane than evaporates from it, and the dominant parameter is the solubility. At later stages, diffusion controls the process (1). Applying Henry's and Fick's laws to the permeation process shows that P can be calculated from the steady state permeation rate across a polymer sheet with the formula:

$$F = \frac{P\,\Delta p}{l} \tag{2}$$

where F = the permeation rate, Δp the pressure difference, l = the thickness of the sheet, and P is the coefficient of permeability (2).

D can be estimated from the permeation rate curve by Ziegel et al.'s (3) modification of Barrer's formula:

$$D = \frac{l^2}{7.2\ t_{1/2}} \tag{3}$$

were $t_{1/2}$ is the time when one half the steady state permeation rate is reached.

These equations were developed to represent the transmission of gases across rubbery polymers, and may not always represent the behavior of the permeability of vapors across glassy polymers (1;4).

In the case of vapors, the relationship between P and permeant molecular weight will depend on how D and S vary with this parameter. D decreases with increasing permeant molecular weight (as molecular size increases) as given by the formula:

$$D = K_d\ V^{-n} \tag{4}$$

where K_d = proportionality constant, V = volume occupied by the molecules as given by V = π a b^2 / 6 where a = length and b = mean root square width. Yi-Yan et al.(5), working with polytetrafluoroethylene (PTFE) and polyfluoroethylenepropylene (PFEP) and hydrocarbon gases from methane to isobutylene as permeants, found that diffusivity decreased as the number of carbons in the chain of the penetrant molecule increased. It was also found that branching had a more marked effect than chain length by itself in decreasing diffusivity. At the same time, for low concentration vapors where interactions between permeant and barrier are negligible, S increases as the boiling point increases (with increasing molecular weight and size), as given by:

$$S = S_0 \, e^{(K_s \, T_b)} \tag{5}$$

where T_b = boiling point (K), S_0 = proportionality constant and K_s is a constant. Combining these equations Zobel (6) found that P will increase or decrease depending on the relative magnitude of the changes in these parameters as given by the expression:

$$\log P = \log K_d + \log S_0 + K_s \, T - n \log V \tag{6}$$

MATERIALS AND METHODS

The homologous series of the ethyl esters were chosen for these experiments because they are present in the natural flavor and aroma of many foods, especially fruits. Polyvinyl alcohol (PVOH) was chosen as the polymer to study because its barrier properties have been shown to be very sensitive to relative humidity, a characteristic shared with the very similar ethylene vinyl alcohol (EVOH), a barrier polymer of growing importance in food packaging.

PVOH (100% hydrolyzed) of approximate molecular weight 115,000 was obtained from Scientific Polymer Products, Inc. A solution of PVOH was made by dissolving the polymer in dimethyl sulfoxide (DMSO) obtained from Fisher Scientific. The diluted solution was cast onto a glass plate with an adapted metal circular wall and evaporated at 65°C until dry. The resulting film was cooled, removed from the glass plate, and degassed at 252 mm Hg and 50°C for 12 hs. The thickness of the films was measured with a Scherr Tumico dial thickness gage model 64-1210-02 on at least 10 different places and the results averaged.

The permeation system consisted of a sample dosing system, a humidifying system, a double sided permeability cell, a detector system, and a mixing system. The permeation apparatus is shown in figure 1. Throughout the system 1/8" copper tubing was used for all connections, except after the humidifying system and the sample dosing system, where 1/8" stainless steel tubing was used.

The permeation cell, obtained from J. Harvey Instruments, consisted of three stainless steel compartments separated by two polymer films. The use of two films at the same time permitted an increase in the sensitivity of the system, by effectively duplicating the permeated area, and consequently, the amount permeated into the downstream side at any time. The combined area was 44.2 cm^2. The permeation cell was kept at a constant temperature in a Fischer Iso-temp 200 series oven. The detector system consisted of a Hewlett-Packard 5790A series gas chromatograph equipped with 5706A dual differential electrometers for dual flame ionization detectors (FID). The detectors response was plotted with a Hewlett-Packard 3390A integrator. A flow of air containing the permeated compound was conducted from the downstream side of the permeation cell into the detector system. This flow was controlled with a fine flow control valve model B-22RS2 from Whitey Co. (valve 5 in figure 1). The sample dosing system consisted of a glass reservoir containing neat sample, and through which a stream of dry air was bubbled to produce a saturated vapor whose concentration was calculated from the

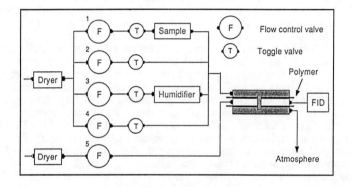

Figure 1.- Permeability system. Schematic diagram.

reported vapor pressure of each ethyl ester. The glass reservoir
was kept at a constant temperature in a refrigerated circulator
water bath (VWR 1145 from Polyscience Corp). The highest
temperature utilized to produce a saturated vapor was 22°C, three
degrees below average room temperature, to prevent condensation of
the sample in the metal tubing. The humidifying system consisted
of a glass reservoir containing distilled water, and through which
a stream of dry air was bubbled to produce a stream of air of
known relative humidity. The glass reservoir was kept at room
temperature. Room temperature at all times was 25°C $\pm$ 2°C. The
mixing system consisted of 4 flow control valves model 8744A from
Brooks Instruments Division of Emerson Electric Co., and 4 toggle
valves model B-OGS2 from Whitey, Co. The incoming dry air was
divided into 4 streams, each controlled by a flow control valve
and a toggle valve. The first stream (valve 1 in figure 1) went
into the sample dosing system. The second stream (valve 2 in
figure 1) diluted the first stream after sample dosing in order to
achieve lower concentrations. The third stream (valve 3 in figure
1) went into the humidifying system and then joined the first two
streams. The fourth stream (valve 4 in figure 1) joined the rest
of the streams and was used to lower the relative humidity or the
sample concentration in the final air mix, which was then
conducted to the upstream side of the permeability cell. All flow
rates were measured with a Hewlett-Packard 0101-0113 soap film
flowmeter and stopwatch. The compressed air was dried by passing
it through a cartridge filled with anhydrous $CaSO_4$.

The films were conditioned by flowing dry air through both
the downstream and the upstream sides of the membranes in the 0%
relative humidity (RH) experiments, or a stream of air of the
desired RH through the downstream side for at least twelve hours
to allow the film to come to equilibrium. For the high relative
humidity tests, the films were conditioned for as long as 48 hs.
After conditioning, a stream of air containing the desired
concentration of ethyl ester and the desired relative humidity was
conducted from the mixing system into the upstream side of the
permeation cell and released into the atmosphere. The ethyl ester
permeated through the films into the center compartment of the
cell (the downstream side) and were carried to the detector. The
permeation rate curve was plotted and the experiment was
terminated when the curve showed no change in 6 hours. Flow rates
were measured again after the completion of the experiment to
ensure that no significant variations had occurred.

RESULTS

The effect of permeant molecular weight on the permeability of
polyvinyl alcohol (PVOH) was determined. Permeation rate
measurements were carried on with five homologous alkyl esters:
ethyl acetate, e. propionate, e. butyrate, e. valerate and e.
caproate. Relative humidity was 0% and the permeant concentration
was 50 μM for all compounds. The film had average thickness of
6.47 x 10^{-2} mm (2.55 mils). Ethyl esters of higher molecular
weights were available, but their low vapor pressures resulted in

permeabilities that were below the detection level of the system. The results are shown in table 1. The permeability coefficients are shown graphically in figure 2. The coefficients of solubility and diffusion are shown in figure 3.

Table 1

Transport coefficients of ethyl esters at 50 μM concentration through 2.55 mils PVOH at 25°C

E. Ester	Mol. Wt.	%[b] S.V.P.	Permeability[a] (moles m/N s)	Diffusion[a] (m^2/s)	Solubility[a] (moles/ N m)
E.Acetate	88.11	0.94	8.64286	16.18500	5.34004
E.Propionate	102.13	2.30	45.30660	21.11090	21.46120
E.Butyrate	116.16	5.58	12.32990	13.39450	9.20520
E.Valerate	130.19	34.24	53.06360	7.76880	68.30350
E.Caproate	144.21	50.20	50.63330 $\times 10^{-17}$	3.13258 $\times 10^{-14}$	161.63500 $\times 10^{-4}$

a. Coefficients are the average of at least two permeation tests.
b. Percentage of Saturation Vapor Pressure.

A second set of experiments determined the effect of relative humidity on the barrier parameters of PVOH. Permeation rates were determined at 0%, 25%, 50% and 75% RH. Ethyl propionate was used at a vapor concentration of 353 μM (16% of saturation) at all RH's. The polymer had an average thickness of 1.78×10^{-2} mm (0.7 mils). The results are shown in table 2. The permeability coefficients are shown graphically in figure 4. Coefficients of diffusion and solubility are shown in figure 5.

Table 2

Transport coefficients of ethyl propionate at 353 μM concentration through 0.7 mils PVOH at 25°C and various relative humidities

Relative Humidity (%)	Permeability[a] (moles m/ N s)	Diffusion[a] (m^2/s)	Solubility[a] (moles/N m)
0	11.57470	19.25730	6.01055
25	11.58410	19.25730	6.01543
50	8.64723	19.77780	4.37219
75	2.43374 $\times 10^{-17}$	9.75705 $\times 10^{-15}$	2.49434 $\times 10^{-3}$

a. Coefficients are the average of at least two permeation tests.

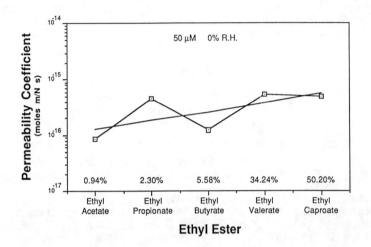

Figure 2.- Ethyl ester permeability in PVOH at 25°C.

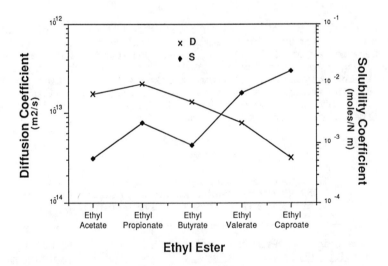

Figure 3.- Ethyl ester diffusivity and solubility in PVOH at 25°C.

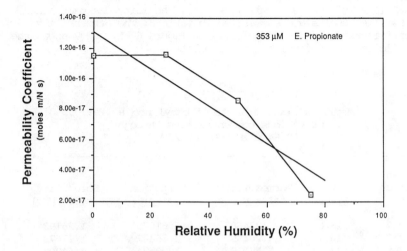

Figure 4.– Permeation of ethyl propionate 353 μM through PVOH at 25°C and various relative humidities.

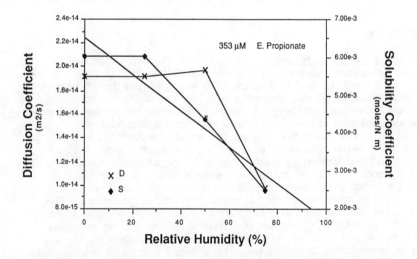

Figure 5.– Diffusivity and solubility of ethyl propionate 353 μM in PVOH at 25°C and various relative humidities.

A third experiment was conducted to determine the effect of ethyl propionate concentration. The average thickness of the films was 2.13 x 10^{-2} mm (0.84 mils). Relative humidity was maintained at 0%. The results are summarized in table 3. The permeability coefficients are shown graphically in figure 6. Figure 7 shows the coefficients of diffusion and solubility.

Table 3

**Transport coefficients of ethyl propionate
at various concentrations through
0.84 mils PVOH at 25°C**

Concen-tration (M)	% S.V.P.	Permeability (moles m/N s)	Diffusion (m^2/s)	Solubility (moles/N m)
297	13.6	3.30468	2.61875	1.26193
353	16.1	3.26189[a]	2.55487[a]	1.27673[a]
705	32.3	3.29019	1.99524	1.64902
929	42.8	4.09684	1.68951	2.42486
1218	56.8	5.58452 x 10^{-17}	1.23235 x 10^{-14}	4.53159 x 10^{-3}

a. Coefficients are the average of at least two permeation tests.

DISCUSSION

Effect of Permeant Molecular Weight. As the molecular weight of the ethyl esters increased, the permeability also increased. Zobel (6), working with the transport of homologous series of permeants through polypropylene, showed a similar trend. Stern et al. (7) measured the permeability of low molecular weight gases and vapors through various silicone polymers and also found higher permeabilities for higher molecular weight permeants, even though most were below molecular weight 60. On the contrary Yi-Yan et al. (5) found decreasing permeability with increasing molecular weight for hydrocarbons with molecular weights below 60 through polytetrafluoroethylene. Similar results were reported by Allen et al. (8) for the permeation of gases through polyacrylonitrile. Zobel (6) stated that for permeants of molecular weight lower than 60, the permeability should decrease with increasing molecular weight, whereas for permeants of weight higher than 60 the opposite occurs. In all cases, the determinant parameter appears to be the solubility of the permeant in the polymer. Zobel (6) also demonstrated that the permeability coefficient will change depending on the relative changes of the diffusion and solubility coefficients. Figure 3 shows that, as molecular weight increased, the diffusion coefficient decreased at a lower rate than the solubility coefficient increased, with the overall result being a

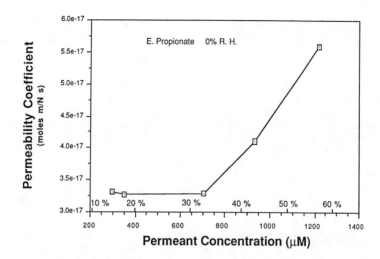

Figure 6.– Permeation of ethyl propionate at various concentrations through PVOH at 25°C.

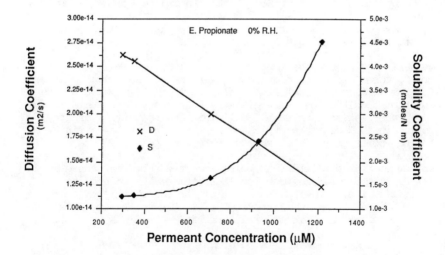

Figure 7.– Diffusion and solubility of ethyl propionate at various concentrations through PVOH at 25°C.

slight increase in permeability (figure 2). This trend has also
been reported by Zobel (6), Stern et al. (7), Allen et al. (8) and
Yi-Yan et al. (5). An explanation for these observations might be
that the diffusion coefficient decreased because the increase in
molecular volume requires the cooperative motion of larger zones
of the polymer chain to open "holes" through which they can
diffuse. This means that a higher energy of activation is required
for the diffusion process to take place, and the probability of an
ester molecule achieving that energy is lower, so the average
speed with which they move through the polymer decreases. The
diffusion values measured for ethyl acetate in these experiments
are close to those given by Zobel (6) for diffusion through
polypropylene, even though a much higher concentration was used
than in our measurements. This demonstrates that PVOH is a better
barrier than polypropylene for alkyl esters.

A good correlation was found between the logarithm of the
diffusion coefficient and the molecular weight of the ethyl esters
in the form:

$$\log D = -11.44974 - 0.01327 \, MW \qquad r = -0.89 \qquad (7)$$

where MW is the molecular weight of the permeant. Our data fit the
equation given by Zobel (6), which relates a permeant's molecular
volume to its diffusivity. Figure 8 shows the relationship between
diffusivity and molecular weight in our experiments. The values
given by equation 7 have been plotted in figure 8 as a continuous
line.

The solubility coefficients increased as molecular weight and
volume increased (figure 3). This trend was also reported by Stern
et al. (7), Yi-Yan et al. (5) and Zobel (6). The solubility
measurements from these experiments for ethyl esters in PVOH are
two to three orders of magnitude lower than those given by Zobel
(6) for various esters through polypropylene and about the same
order of magnitude as those given by DeLassus et al. (this volume)
for trans-2-hexenal through EVOH. This low solubility of ethyl
esters in PVOH might be explained by the relative low affinity of
the non-polar ethyl ester molecules for the polar PVOH matrix. The
polymer represents a lipophobic medium into which the ethyl ester
molecules have difficulty in being absorbed. However, there is a
linear relationship between the logarithm of the solubility
coefficients and the boiling point of the permeant (9;6) as given
in equation 5. This equation, applied to the solubility data from
these experiments gives the relationship:

$$\log S = -8.62227 + 0.01527 \, T \qquad r = 0.90 \qquad (8)$$

where T is the boiling point of the permeant in K. These results
are shown in figure 9, where the values predicted by equation 8
are plotted as a continuous line. This explains the upward trend
shown by the solubility coefficients in figure 3. Combining
equations 7 and 8 for solubility and diffusivity, the permeability
of any ethyl ester through PVOH can be calculated with:

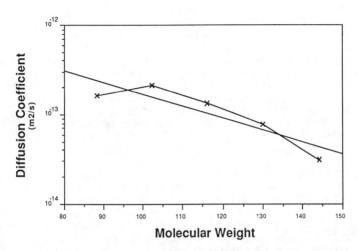

Figure 8.- Relationship between permeant molecular weight and its diffusivity in PVOH at 25 °C.

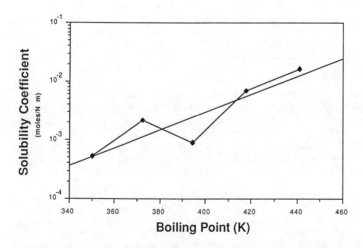

Figure 9.- Relationship between permeant boiling point and solubility in PVOH at 25°C.

$$\log P = -20.07201 - 0.01327 \text{ MW} + 0.05127 \text{ T} \qquad (9)$$

a modification of the equation given by Zobel (6). This equation combines the changes in diffusivity and solubility given by equations 7 and 8, and predicts how the permeability will change. Equation 9 is shown in figure 2 as a continuous line. Another explanation for increasing solubility of the ethyl esters as the molecular weight increased is that, in order to test all permeants at the same concentrations, the high molecular weight esters had to be tested at vapor pressures that were closer to their saturation vapor pressure than those for the low molecular weight esters. As will be described in a later section, ethyl esters show a marked concentration effect at concentrations higher than 30% of the saturation vapor pressure, and this results in higher permeabilities.

Since the molecular weight of all esters was larger than 60, as the molecular weight increases, the contribution of the increasing solubility coefficient outweighs the contribution of the decreasing diffusion coefficient, resulting in an increasing permeability for larger molecules, i.e., even though the molecules are moving slower through the polymer, a larger number of them are moving, and more emerge at the other side of the barrier.

Effect of Relative Humidity. For reasons that will be discussed in the next section, it was considered that the permeant concentrations at which the tests were run did not cause plasticization in the polymer.

Figure 4 shows a trend towards lower permeability values as the relative humidity increased. This was contrary to what was expected. Theory predicted that the plasticizing effect of water vapor on an hydrophilic polymer like PVOH would increase the permeability coefficient by increasing the diffusivity due to the higher mobility acquired by the polymer network. Results by Ito (10) showed that this holds for the permeation of CO_2 through PVOH. Work by DeLassus et al. (this volume) on transport of trans–2–hexenal trough EVOH, a polymer similar to PVOH, also agreed with theory and showed an increased permeability at higher relative humidities. Data by Hatzidimitriu et al. (11) for the transport of various organic vapors through various multilayered films showed that for some of them the permeability increases at higher relative humidities, while for others the opposite occurred.

Analysis of diffusion and solubility data from our experiments, (figure 5), explains why the permeability decreased as relative humidity increased. Diffusion coefficients remained constant and showed a slight downward trend only at high relative humidities. Solubility coefficients showed a clear trend towards lower solubilities at higher relative humidities. Thus, the decrease in permeability was mainly a result of the lowering of the solubility as relative humidity increased.

The decrease in solubility may be explained by a competition effect between the water molecules and the ethyl ester molecules. The highly polar water molecules were able to compete more effectively than the non-polar ethyl ester molecules for the absorption sites of the also polar PVOH, establishing a solubility partition coefficient unfavorable for the organic permeant. If the solubility of the permeant in the polymer is lowered, when all other parameters are constant, then the permeability will also decrease. The same idea was suggested by DeLassus et al. (this volume) for the permeation of trans-2-hexenal through EVOH below its T_g.

The presence of water molecules in the polymer matrix, occupying holes that otherwise would be available for the diffusion of permeant molecules, effectively increased the length of the viable diffusion paths, diminishing the diffusivity. Furthermore, water molecules present throughout the polymer represented a lipophobic barrier to non-polar molecules. These two phenomena, combined with the fact that ethyl esters are immiscible with water, seemed to increase the resistance to diffusion just enough to counteract the relaxation of the polymer network brought about by the plasticization with water.

Regression analysis of the relationship between permeability and solubility coefficients with relative humidity presented in figures 4 and 5, showed these relationships to be linear, following the forms:

$$S = S_0 + K_S \ RH \tag{10}$$

and

$$P = P_0 + K_P \ RH \tag{11}$$

where S_0 and P_0 are the solubility and permeability coefficients of the barrier at 0% relative humidity, K_S and K_P are constants, and RH is the relative humidity. The parameters for equations 10 and 11 are shown in table 4.

Table 4
Relationship between permeability and diffusivity coefficients and relative humidity.
Linear equation parameters

E.Ester	P_0	K_P	r	S_0	K_S	r
E.Propionate	1.311×10^{-16}	-1.214×10^{-18}	-0.91	6.552×10^{-3}	-4.887×10^{-5}	-0.94

P_0 and S_0 in moles m / N s

Effect of Permeant Concentration. Figure 6 shows that the permeability appeared linear at low concentrations and low percentages of the saturation vapor pressure. At concentrations greater than 30% of saturation, an observable concentration effect was seen, and the permeability increased non-linearly. Zobel (6) found the same phenomenon for the permeation of d-limonene through polypropylene, although the onset of non-linearity occurred around 15% of saturation vapor pressure. The fact that ethyl propionate and polyvinyl alcohol are more chemically dissimilar than limonene and polypropylene, may explain why the non-lineal concentration effect appeared at a higher percentage of saturation vapor pressure for the ethyl propionate / PVOH pair.

The fact that the permeability of organic vapors at low vapor concentrations follows a linear relationship supports Zobel's (6) claim that the transport of these compounds at low concentrations follows Fickian behavior. Since most of the experiments of this work were performed at vapor pressures well below saturation (except where otherwise noted), this fact also validates the use of equations 1, 2 and 3, derived from assumptions of Fickian behavior, to calculate the coefficients of permeability diffusivity and solubility.

The diffusion coefficient decreased linearly as the concentration increased, while at the same time, the solubility coefficient first increased linearly and, at 30% saturation vapor pressure, increased non-linearly as the concentration increased. The combination of the linearly increasing solubility and the linearly decreasing diffusivity at low permeant concentrations combined to result in the linear portion of the permeability curve (figure 6). At higher concentrations the solubility increased much more rapidly than the diffusion decreased, and this combination resulted in an increasing permeability at higher concentrations.

The fact that the solubility coefficient increased non-linearly at high concentrations suggests that the polymer became plasticized. Theory predicted that a plasticized polymer would have increased permeability, and our results conform to theory. Similar results were found by Zobel (6) for d-limonene through polypropylene, and by Baner et al. (12) for toluene through polypropylene and Saran (PVDC). On the contrary, DeLassus et al. (this volume) reported finding no permeant concentration effect for the transport of trans-2-hexenal through LDPE. Theory predicted (13;9) that the increased permeability would be a result of both an increase in solubility and an increase in diffusivity. Our results conformed to theory in showing an increased solubility, but also showed that, in the case of diffusivity, exactly the opposite phenomenon occurred than what was predicted (figure 7).

CONCLUSIONS

The permeability of PVOH to alkyl esters is extremely low. The higher molecular weight alkyl esters are more soluble and more permeable than the lower molecular weight esters. This may mean that the polymer will absorb and transport some flavor components more than others. In this way, given long storage times, the food would no longer contain the same proportions of each compound it originally had. A "flavor imbalance" could develop and be detectable when consumed. This problem may be present when EVOH is utilized in retort applications, as well as in polymer packaged fruit concentrates and juices.

Higher relative humidities enhanced the barrier properties of PVOH against low molecular weight alkyl esters. This appears to result from lipophobicities of the polymer and the water. It is likely that the amount of water in the polymer is more important for this effect, than the relative humidity of the atmosphere. The study of the effect of ethyl ester concentration on its permeability through PVOH showed a strong concentration effect at permeant vapor pressures higher than 30% of saturation vapor pressure. This indicates that the results of permeation tests performed at saturation vapor pressures will not be directly applicable to food packaging situations. Permeabilities measured at saturation vapor pressures will lie in the non-linear part of the permeation curve, whereas the vapor pressure of the food components likely to permeate will be very low, and their permeability will lie in the linear part of the permeation curve, where transport behaves in a Fickian manner. Materials reported as inappropriate for food packaging applications on the basis of permeability tests at saturation vapor pressures may actually offer sufficient protection for food flavors and aromas. However, these results also confirm the tendency of higher molecular weight esters to permeate more readily than low molecular weight esters, with the consequent result of a food with modified flavor character, where the high molecular weight aromas have been diminished while the low molecular weight components have suffered little or no change.

Further studies should be performed to determine if the onset of non-linearity occurs at the same point for all ethyl esters. It is conceivable that high molecular weight esters, with their higher solubilities, may plasticize the membrane at lower vapor pressures, causing a non-linear concentration effect at lower percentages of the saturation vapor pressure. Our work is currently attempting to develop predictive equations for the loss of aroma quality in given food package applications.

LITERATURE CITED

1. Lomax, M. Polymer Testing. Vol. 1, No. 2, 1980. pp. 105–147.
2. Stannett, V. Journal of Membrane Science. Vol.3, No. 2–4, 1978. pp. 97–115.
3. Ziegel, K. D.; Frensdorff, H. K. and Blair, D.E. Journal of Polymer Science: Part A–2. Vol. 7, No. 5, 1969. pp. 809–819.
4. Zobel, M. G. R. Polymer Testing. Vol. 3, No. 2, 1982. pp. 133–142.
5. Yi-Yan, N.; Felder, R. M. and Koros, W. J. Journal of Applied Polymer Science. Vol. 25, 1980. pp. 1755–1774.
6. Zobel, M. G. R. Polymer Testing. Vol. 5, No. 2, 1985. pp. 153–165.
7. Stern, S. A.; Shah, V. M.; and Hardy, B. J. Journal of Polymer Science: Part B: Polymer Physics. Vol. 25, No. 6, 1987. pp. 1263–1298.

8. Allen, S. M.; Fujii, M.; Stannett, V. Hopfenberg, H. B.; and Williams, J. L. Journal of Membrane Science. Vol. 2, 1977. pp. 153-164.
9. Stannett, V. Diffusion in Polymers. J. Crank and G. S. Park Editors. 1968. Academic Press, New York. pp. 41-73.
10. Ito, Y. Kobunshi kagaku. Vol. 18, 1961. pp: 158-162.
11. Hatzidimitriu, E.; Gilbert, S. G. and Loukakis, G. Journal of Food Science. Vol. 52, No. 2, 1987. pp. 472-474.
12. Baner, A. L.; Hernandez, R. J.; Jayaraman, K.; and Giacin, J. R. Current Technologies in Flexible Packaging. ASTM STP 912. pp. 49-62.
13. Lebovits, A. Modern Plastics. Vol. 43, No. 7, March 1966. pp. 139+.

RECEIVED December 11, 1987

Chapter 5

Flavor–Polymer Interactions

Coffee Aroma Alteration

Ann L. Hriciga and Donald J. Stadelman

Polymer Products Department, E. I. du Pont de Nemours and Company, Experimental Station, Wilmington, DE 19898

Because organoleptic testing is time-consuming and expensive, we have developed a gas chromatographic test to screen polymer films for their ability to change the aroma profile of coffee. The retention or alteration of flavor components of foods packaged in contact with polymer films is of concern both to suppliers of those polymers and to food manufacturers. Ground coffee is a product which is beginning to move away from the traditional metal can and into new, polymeric types of packaging. Incubation of ground coffee with films in sealed vials for up to three days preceded gas chromatographic analysis of volatile compounds in the headspace. After statistically analyzing the data, we ranked the films for alterations to the coffee aroma profile. Our ranking matched that of a professional olfactory panel working with ground coffee stored for six months in bags made from the same polymers. Our method results in a substantial savings in time and cost with no change in qualitative result.

The recent trend on the part of the food industry to replace traditional glass and metal containers with plastic ones has focussed attention on the interactions between aromas or flavors and the polymers used in these new structures. In order for these plastic packages to gain widespread acceptance by the consumer, foods packaged in plastics must continue to smell and taste as they did in their former packaging.

Polymers may interfere with those aromas in several ways. Adsorption or absorption by the plastic may simply reduce the total volatile content of the packaged material, resulting in a loss of odor. Alternatively, the polymer may alter the characteristic aroma of a food by selectively absorbing one or more key compounds which make up the characteristic aroma of a food. Off-smells may be produced if the polymer acts to chemically change a component of the food: enhanced degradation is an example of this. Polymeric materials themselves can contribute aromas from such things as residual monomers, solvents or processing additives.

The most reliable method of evaluating a polymer's impact on a packaged food is sensory evaluation of the food by a human panel. A container must be fabricated, the food sealed inside and the entire package incubated to simulate shelf storage before panel testing can be done. This process may cost thousands of dollars for each polymer and food tested. In addition, the incubation period builds an inherent delay into the evaluation process. The number of samples that can be evaluated for aroma by a single panel limits throughput.

The development of acceptable plastics for food containers could be quickened if an alternate sensory evaluation test could be designed. Many more materials can be tested with a relatively short, inexpensive, analytical method. Such a test does not replace or eliminate sensory evaluation, but rather acts as a prescreening so that only the most promising plastics are used in candidate packages. These packages would then undergo the same rigorous aroma/flavor evaluation that is now being used prior to consumer marketing.

The goal of our work was to establish such a test. We had to choose a source of flavors (real food versus odorous compounds), some type of packaging material and construction, and an analytical technique. Any method developed had to be validated by comparing its results to those from a sensory evaluation of the same flavor-package combination.

To simplify our test, we decided to start by looking at alterations in aroma, leaving taste for future study. Our choice of technique then became rather straightforward – anything which is volatile enough to be smelled is volatile enough to be analyzed by gas chromatography (GC). We recognize the inherent limitations of this technique, though. A gas chromatograph cannot detect all aroma components with the same sensitivity as the human nose. Consequently, our technique may not see alterations in concentrations of a component which is key to the aroma, but present only at an extremely low level. Another benefit of GC is that addition of components from the polymer, such as residual solvents, can also be detected although, again, extremely low levels may either be below the detection limit of the GC or be obscured by components of the aroma.

In order to make our system as realistic as possible, we decided not to use model compounds for development of this method, but rather a real food. Actual food aromas are often very complex mixtures and it is not necessarily possible to predict which components of the mixture will be most affected by the polymeric packaging material. However, a strong food aroma often has a high concentration of volatile components which can then be analyzed by GC. Thus, we will be more apt to detect changes in this well-defined aroma profile. The ideal material is a food whose strong aroma is positively perceived by the consumer.

This thought process made the choice of food relatively straightforward. One food whose aroma is very important to its consumer appeal is ground coffee. To most coffee drinkers, the smell of a newly opened container of coffee is pure perfume. If coffee is packaged in a way that alters that aroma significantly, the consumer will not buy it more than once. In the past five years, we have seen the introduction of new types of coffee packages on supermarket shelves, so we know the coffee industry is evaluating new packaging materials. Consequently, ground coffee seemed a very good choice as the food to be evaluated. As an added benefit, it is a solid material and is easily handled.

With the choice of coffee as food and the knowledge that the new coffee packages are flexible, it was natural to choose polymer films as the materials to evaluate. Two DuPont films (Film A and Film B) had already been evaluated by a sensory panel for their impact on coffee aroma; they became the first materials we examined. Indeed, the results of that panel provided us with the means of comparing our laboratory method with a real sensory evaluation.

Methodology

The panel tested ground coffee which had been stored for six months in pouches made from polymer film/aluminum foil laminates. The aluminum foil outer layer acted as a moisture, oxygen and light barrier so that only the interaction between the coffee and the polymeric inner layer was being examined by the panel. The control was simply ground coffee packed in a metal can. In this test, coffee packed in the metal can had the best aroma of the three cases while coffee packed with Film A had an acceptable aroma. The aroma of the coffee packed with Film B was unacceptable.

Simply put, our test involves evaluating the chromatographic profile of coffee aroma in the presence and absence of polymer films. This is a comparative technique which establishes a ranking of the films tested, from most similar to most dissimilar to a sensorially acceptable container.

Because we didn't know ahead of time what the magnitude of the alterations in the aroma would be, we used a very high ratio of polymer film to ground coffee to increase our chances of seeing changes in the aroma profile. Although this is not the situation in a real flexible coffee package, the high ratio maximizes the interaction between the polymer and the coffee and should make any differences in the aroma more pronounced.

We incubated roughly five grams of ground coffee (from freshly opened cans of ground coffee) either alone or in the presence of a film candidate. Films of the same thickness (~1.8 mil) and area (4" x 5") were used so that both the volume and the surface area are the same from sample to sample. The coffee, with or without film, was placed into a glass container (approximately 40 mL volume) which was then sealed with a Teflon®-faced septum and cap. This sealing system allowed us to periodically sample the headspace of the vial with a gas-tight syringe. Incubation was at 500°C for a period of up to three days. The high temperature was chosen to speed up the aging process and as such it did not replicate storage at room temperature. (However, it is a temperature which packaged coffee might see inside unrefrigerated vehicles while being transported in the southern portions of the U. S. during the summer.)

Analysis of the headspace, which is simply the aroma of the coffee or the coffee plus polymer system, was done by gas chromatography. Previous work on the analysis of coffee aroma by several workers (1,2) used narrow bore glass capillary columns with polar

stationary phases. Most of the volatiles eluted above 400°C in
these studies. Consequently, we chose a 25 meter, 0.53 mm I. D.
fused silica capillary column coated with three microns of Carbowax
20M (Quadrex Corp.). With this column, there was no need to split
or concentrate the aliquot of the headspace which we wished to
chromatograph and we avoided problems with possible discrimination
or loss of volatile compounds.

The carrier gas was helium at 8 mL/min. The injection port
was maintained at 200°C and the flame ionization detector was kept
at 250°C. The oven temperature started at 500°C and was programmed
to 250°C at 800/min and then held. An airtight syringe was used to
inject 0.5 mL of the headspace onto the column. In order to
maintain a constant pressure in the sample vial, 0.5 mL of
laboratory air was injected into the vial using the same syringe
prior to each sampling.

For the development of this method, all the samples (Films A
and B) and the control (coffee alone) were prepared in triplicate.
Each vial was sampled once a day for three days.

Results

As can be seen from the chromatogram (Fig. 1), the aroma profile is
quite complicated. In fact, well over 150 different compounds have
been identified in the literature (2,3) as volatile components of
coffee. To start characterizing the aroma profile, we chose the 49
largest components in the chromatograms of the control headspace
(coffee alone) and followed their behavior as a function of time in
all nine vials. The peak area of each of these components was
divided by the total area of these components, and this ratio was
plotted as a function of incubation time.

The behavior of each component was characterized by fitting
the peak area ratios to a curve of either zero, first or second
order in time and generating values which represented the expected
upper and lower limits (at the 95% confidence level) of each curve.
Each component, therefore, had three curves associated with its
behavior – one in glass alone, one in the presence of Film A and
one in the presence of Film B. The curves for a given component in
these three systems were compared to one another and declared to be
statistically different from one another if there was no overlap
between the curves. Typical results are discussed below.

In one common result, the component behaved quantitatively the
same in the three systems. The presence of the films made no
difference in the behavior of that particular component, so any
change in the aroma when the films were present was not due to
absorption of this component.

A second case is one in which the component behaved the same
in the presence of either Film A or Film B, but differently when in
glass alone. Although components which fell into this category
might contribute to the overall aroma of ground coffee, they were
not changed to an unacceptable level (which was the same in the
presence of either film) because coffee packaged with Film A passed
sensory evaluation. These components cannot be used to rank the
films because they react the same to either film.

The category of result where a given component behaved
statistically differently in the presence of Film A than it did
with Film B comprised 23 components of the 49 which we examined.
This category could be further divided into three subclasses:

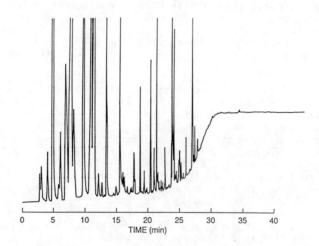

Figure 1:   This is a typical chromatogram of the aroma of freshly ground coffee. The rise in the baseline after 24 minutes is due to the temperature programming of the column itself and is not due to components in the aroma.

o The component behaved the same in glass alone and with Film
  A, but differently with Film B (13 components). See Fig. 2.
o The component behaved the same in glass alone and with Film
  B, but differently with Film A (no component in the headspace
  fell into this subclass).
o The component behaved differently in each of the three
  systems of incubation (10 components).

This last subclass can be further divided by ranking the component's behavior in the three systems by relative concentration.
For seven components, the component concentration was highest in
glass alone, next highest with Film A, and lowest with Film B
(Fig. 3). Only two components had the highest concentration in
glass alone, next highest with Film B, and lowest in Film A. The
remaining component in this subclass had a concentration in glass
alone that was intermediate between the values with Film A and Film
B.

A review of these results yields the following information.
In 20 out of 23 components (87%) where behavior in the presence of
Film A was different from that with Film B, Film A altered the
component's behavior less than did Film B. In only three cases was
the situation different. We had been hoping to find several components which would show behavior for Film A that was intermediate
between no film and Film B; what we found was that in 87% of the
cases where there was a difference, the difference was in exactly
that direction. We knew from the sensory panel testing that coffee
packaged with Film A had an acceptable aroma, although not as good
as that of coffee packaged in a traditional metal can. Coffee
packaged with Film B, however, had an unacceptable aroma. So here
we have a first ranking of packaging materials: Film B is
unacceptable, Film A is acceptable, but the metal can is best of
all. Clearly our GC results echo the same pattern.

Future Work

How can this method be used and improved? First, we have found 20
peaks which seem to correlate with the sensory analysis. We can
reduce the number of components we track; this will reduce the time
spent on statistical analysis. Secondly, we can try to identify
these 20 peaks. We might be able to assemble a synthetic mixture
of aroma components to be used in this test, reducing the
complexity of the chromatograms and providing us with a standard
mixture which we can control. Such a mixture would eliminate the
lot-to-lot variation which may be present in the aromas of
commercially obtained packaged coffee, and we could directly
compare experiments done with different films at different times.
Currently, we can only rank films which were tested with the same
lot and age of coffee.

Identification of these peaks, which we are working on by
GC/MS, along with some knowledge of the chemical nature of these
polymer films, should give us a better understanding of how the
aroma is interacting with the films. We would have more
information on how to design or choose candidate packaging
material.

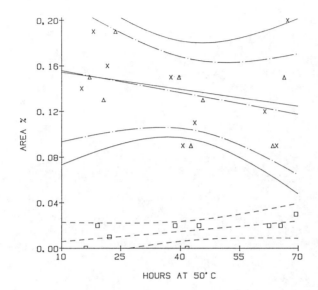

HOURS AT 50°C

Figure 2:   The behavior of Component 46 is plotted as a
function of time for each of the three systems.   Three lines
are associated with each system: the middle line of a set is
the best fit to the data and the upper and lower lines are the
95% confidence level limits associated with that fit.   Symbols
used are as follows: x, ——— Coffee alone; Δ, —·— Coffee
with Film A; □, ----- Coffee with Film B.   Component 46
exhibits the same behavior in the presence of the glass vial
alone as it does in the presence of Film A.   Film B, however,
reduces the relative concentration of this component in the
headspace by at least a factor of four.

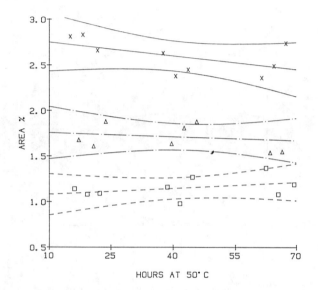

Figure 3:   The behavior of Component 26 is plotted as a
function of time for each of the three systems.   Three lines
are associated with each system: the middle line of a set is
the best fit to the data and the upper and lower lines are the
95% confidence level limits associated with that fit.   Symbols
used are as follows: x, ──── Coffee alone; Δ, ──·── Coffee
with Film A; □, ----- Coffee with Film B.   Component 26 behaves
differently in each of the three cases, with its highest
relative concentration in the glass vial alone.   Film A reduces
this component's level to 75% of the relative concentration in
the glass vial alone, but Film B has an even greater effect,
diminishing Component 26 to 40% of its relative concentration
in the glass vial alone.

Acknowledgment

The help of statistician Mary Dunleavy is gratefully acknowledged.

References

1. Vitzthum, O. G.; Werkhoff, P. In Analysis of Food and
   Beverages; Charalambous, G., Ed.; Academic:New York, 1978;
   pp 115–133.

2. Tressl, R.; Silwar, R. Agric. Food Chem. 1981, 29,
   1078–1082.

3. Stoffelsma, J.; Sipma, G.; Kettenes, D. K.; Pypker, J.
   J. Agric. Food Chem. 1968, 1000–1004.

RECEIVED September 24, 1987

Chapter 6

Interaction Between Aseptically Filled Citrus Products and Laminated Structures

C. H. Mannheim [1], J. Miltz [1,2], and N. Passy [1]

[1]Department of Food Engineering and Biotechnology, Technion–Israel Institute of Technology, Haifa, 32000, Israel

The interaction between polyethylene(PE) contact surfaces of carton packs and "Bag in the box" packages with citrus products and model solutions was evaluated. The presence of corona treated polyethylene films in model solution and juices accelerated ascorbic acid de-gradation and browning. Untreated PE films did not accelerate these reactions. In all cases d-limonene concentration in juices in contact with PE surfaces was reduced. In both orange and grape fruit juices, aseptically filled into cartons the extent of browning and loss of ascorbic acid was greater than in same juices stored in glass jars. Sensory evaluations showed a significant difference between juices stored in carton packs and in glass jars, at ambient temp. after 10-12 weeks. The cartons were found not to be completely gas tight. Storage life data of orange juice in bag in the box packages are presented. Dynamic vibration tests of bag in the box packaging systems, to simulate transport distribution environment, were carried out. In these tests behaviour of different metallized structures, to flex-cracking and pinholing when in contact with citrus products and simulants was evaluated.

Aseptic packaging has been commercially viable in Europe for several decades. In the U.S.A. aseptic packaging has become one of the fastest growing areas of food packaging since the Food and Drug Administration (FDA) approval,in 1981, of the use of hydrogen peroxide for sterilizing polyethylene (PE) contact surfaces (1). Approval was later extended to include olefin polymers, EVA copolymers and polyethylene phthalate polymers (2).

Fruit juices in general and citrus juices in particular are at present extensively packed aseptically (3). Most of these products are packed today in semi-rigid brick shaped laminated

[2]Current address: Pillsbury Company, Research and Development Laboratories, Minneapolis, MN 55402

0097–6156/88/0365–0068$06.00/0
© 1988 American Chemical Society

PE/carboard/ aluminum foil/PE packages. Another popular packaging system for juices and concentrates is the "bag-in-the-box" ranging in size from 2 to 1200 liters. The structure of the bag is generally a laminate made of PE/metalized polyester/PE with an inner PE liner. The metallized polyester may be replaced by a non metallized high barrier film. A rigid coextruded multilayer high barrier container made on form-fill-seal lines offers another choice of packaging for above products (4).

Factors Affecting Shelf-Life of Aseptically Packed Citrus Products

The shelf-life of pasteurized, hot or aseptically filled, juices or concentrates is limited by chemical reactions which are primarily influenced by storage temperature, as well as oxygen and light. In plastic laminates, compared with glass, gas permeation of the packaging material as well as absorption by the contact surface or migration of low molecular weight compounds may influence shelf-life (5,6).

In storage studies of orange juice in Tetra Brick and glass bottles, Durr et. al (7) reconfirmed that storage temperature was the main parameter affecting shelf-life of orange juice. In addition they found that orange oil, expressed as d-limonene, was absorbed by the polyethylene contact layer of the Tetra Pack. These authors claimed that this could be considered an advantage since limonene is a precursor to off-flavor components. Marshall et.al (8) also reported absorption of d-limonene from orange juice into polyolefins. They found that the loss of d-limonene into the contact layer was directly related to the thickness of the polyolefin layer and not its oxygen permeability. Marshal et.al(8) also quantified the absorption of several terpenes and other flavorings from orange juice into low density polyethylene (LDPE). They found that the predominant compounds absorbed were the terpenes and sesquiterpenes. More than 60% of the d-limonene in the juice was absorbed by LDPE but only 45% by Surlyn. Other terpenes such as pinene and myrcene were absorbed to greater extents. Infra red spectra of LDPE showed that the terpenes were absorbed rather than adsorbed. Thus since many desirable flavor components in citrus juices are oil soluble, their loss by absorbtion as mentioned above can adversly affect flavor of juices. Furthermore, Scanning Electron Microscope (SEM) pictures of LDPE after exposure to orange juice, showed significant swelling of the polymer, inferring severe localized internal stresses in the polymer. These stresses indicated a possible cause for delamination problems occasionally encountered in LDPE laminated board structures. The adhesion between LDPE and foil was much less than between Surlyn and foil. The presence of d-limonene in the polymer aided in the absorption of other compounds, such as carotenoid pigments, thus reducing color of juice (8). Buchner (9) also mentions reduction in aroma bearing orange oil into the polyethylene contact layer of the three piece HYPA S aseptic package.

Gherardi (10) and Granzer (11) also investigated quality changes of fruit juices in carton packs as compared with glass packages. Both found, separately, that juices and nectars

deteriorated faster in the carton packs than in glass bottles. In
products with low fruit content, such as fruit drinks, the
difference between packages was small. Gherardi (10), found high
oxygen and low carbon dioxide contents in the head space of carton
packs versus low oxygen and high carbon dioxide in glass bottles.
The CO_2 content was attributed to ascorbic acid degradation.
Therefore, Gherardi (10) concluded that the main cause for the
difference in keeping quality between products in carton packs and
glass was due to higher oxygen transfer rates into the former
package. Granzer (11) also claimed that the oxygen permeability
of carton packs caused the faster deterioration of juices packed
in them. Granzer (11) stated, without showing proof, that the
source of the oxygen permeation into the carton pack was the 22cm
long side seam. It is known that since then Tetra Pack improved
the side seam by adding a strip of PET at this place.

Koszinowski and Piringer (12) investigated the origin of
off-flavors in various packages. They established gas
chromatographic and mass spectra procedures coupled with sensory
threshold odor analysis for the determination of off-flavor
emanation from packaging materials. They found reaction products
of raw materials as well as a combination of oxidation and
pyrolytic elimination reactions in polyethylene or paperboard to
be responsible for off-odors.

Products packed in "bag-in-the-box" packages, especially in the
larger sizes, apparently also have a reduced shelf life as
compared to those packed in conventional metallic packages, due
mainly to damage caused to the films during transportation
(13). The reduced shelf life in these packaging systems is
probably due to flex cracking which causes pinholes and even
"windows" in the metallized laminate thus reducing the barrier
properties.
     In our laboratory the interaction between polyethylene contact
surfaces of carton laminates and "bag-in- the box" structures and
citrus products are being investigated. Some results of these
studies are presented here.

Evaluation of Aseptic Packages and Citrus Products -
Experimental Results

Carton Packages. Orange and grapefruit juices were packed
aseptically on a commercial line into cartons and glass jars (14).
The carton structure was a laminate of polyethylene/carton/
aluminum/ polethylene made by PKL- Combibloc, Germany. The carton
dimensions were 10x6x17 cm. Browning and ascorbic acid degradation
 of orange juices are presented in Figures 1 and 2. In all cases
the rates of the deteriorative reactions were higher in cartons as
compared to glass. Differences in browning and ascorbic acid
degradation between juices in glass and cartons were initially
small and increased with storage time. Initially, juices in glass
and cartons had similar dissolved and headspace oxygen values
which affected the above reactions in the same fashion. Later on,

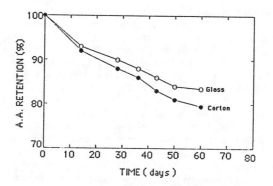

Figure 1. Ascorbic acid retention in orange juice aseptically packed into glass jars and carton packs, stored at 25°C. (Reproduced with permission from Ref. 14. Copyright 1987 Inst. of Food Technologists.)

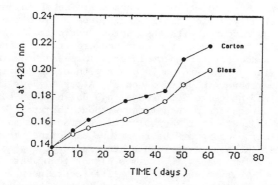

Figure 2. Browning of aseptically filled orange juice in glass jars and carton packs, stored at 25°C. (Reproduced with permission from Ref. 14. Copyright 1987 Inst. of Food Technologists.)

the effect of the laminated package became more pronounced due to its oxygen transmissibility as well as due to product-package interaction. Similar results (not shown) were obtained with grapefruit juice.

A rapid loss of d-limonene, from 70 to 50mg/l for orange juice was observed in carton packs with juices during the first days of ambient (25°C) storage (Figure 3). Similar results were obtained by others (7,8) who claimed absorption by the film as the cause for this phenomenon. It can be compensated for, if required, by increasing the initial d-limonene concentration. However, d-limonene oxidation may also occur and is a subject for further studies.

Triangle taste comparisons showed that, for orange juice after about 2.5 months storage at 25°C, there was a significant (p< 0.05) difference between juices in cartons and glass containers (Fig. 4). In grapefruit juices a significant taste difference (p< 0.05) was found after about 3 months (Fig. 4). The longer period to obtain a taste difference in grapefruit juice is due to the stronger flavor of this juice which tends to mask small flavor changes. Color differences were noted easier in grapefruit juices than in orange juices. In orange juice, some carotenoids may be oxidized and have a bleaching effect which compensates for the change in appearance due to browning, thus explaining the color difference between the two juices (15).

Carton Strips. The interaction between carton packs and juices was evaluated by immersing strips of carton laminates in juices stored in hermetically sealed glass jars. Surface to volume ratios of 4:1 and 6:1, as compared with the actual ratio in 1 liter packs. were used. The contact of orange and grapefruit juices with the carton strips accelerated ascorbic acid loss and browning as compared to similar juices stored in glass without the laminate, at the same temperatures (Figures 5 and 6). After 14 days there was also a significant taste difference (p<0.05) between samples in contact with carton strips and those without the strips. These results seem to indicate that the polymeric surface had an accelerating effect on some of the reactions affecting shelf-life of citrus juices. In these experiments, the ratio of contacting surface to juice volume was 4 to 6 times higher than that in 1 liter containers. This means that in 1 liter containers the effect will be somewhat delayed as compared to the above test, but in smaller packages (i.e. 250 ml) the shelf life would be shorter due to this effect.

Polyethylene films. In order to elucidate the effect of the polyethylene contact surface, and separate any possible contribution of the carton itself, untreated and corona treated (oxidized) polyethylene strips, made of LDPE of similar properties to that used in PKL cartons, were immersed in model solutions. The model solution consisted of 10% sucrose, 0.5% citric acid, 0.1% ascorbic acid, 0.05% emulsified commercial orange oil and 0.1% potassium sorbate. The effect of the untreated and corona-treated

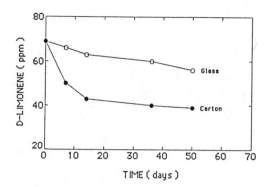

Figure 3. D-limonene concentration in orange juice aseptically filled into glass jars and carton packs, stored at 25°C. (Reproduced with permission from Ref. 14. Copyright 1987 Inst. of Food Technologists.)

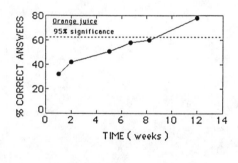

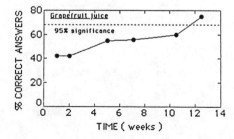

Figure 4. Triangle taste comparisons between juices aseptically filled into glass jars and carton packs, stored at 25°C.

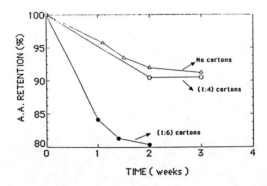

Figure 5. Ascorbic acid retention of orange juice stored at 35°C as effected by contact with carton strips.

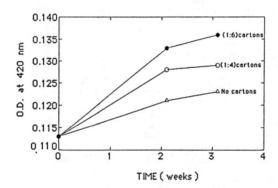

Figure 6. Browning of orange juice stored at 25°C as affected by contact with carton strips.

LDPE strips on ascorbic acid degradation is shown in Figure 7. From these results it becomes clear that the polethylene contact surface accelerated the rate of ascorbic acid degradation with the oxidized film having a greater effect. In the model solutions a rapid reduction (40-60% within about six hours) in d-limonene content took place in samples containing LDPE strips as compared to a loss of 10% in the blanks (Figure. 8). Absorption of d-limonene by the LDPE accounted for this loss. In a triangle taste comparison of water with and without LDPE strips, stored in sealed glass jars at 35°C for 48 hours, all 12 tasters were able to distinguish between the samples, indicating the presence of a strong off-flavor in the water which had been in contact with LDPE strips.

Strips of LDPE, from the inner liner of the bag in the box at a surface to volume ratio double the actual ratio in a 6 gallon bag, were inserted into glass jars, which were hot filled with orange juice and stored at 35°C. In contrast to the degradative effect of the PE contact layer of the cartons, no detrimental effect on browning, ascorbic acid and taste, as compared to juice in glass, was found in this case. However, d-limonene content was reduced to about half the initial content in a short time also in this case.

These differences may be accounted for by the different manufacturing techniques of the polyethylene contact surfaces. A high extrusion temperature, followed by corona treatment which causes oxidation, was used for the PE layer of the cartons, whereas the PE liner for the bag-in-the-box is a LDPE extruded at a lower temperature and without corona treatment. This indicates that an oxidized polymer surface may accelerate reactions resulting in a reduced shelf-life. These results were confirmed by (6) who presented the effect of high temperature extrusion lamination of an Ionomer on development of off-flavors.

Bag-In-The-Box. Shelf-life of aseptically filled orange juice, in bag-in-the box made from various structures of laminated polymers, metallized and unmetallized, was evaluated. After filling with juice, the bags were flushed with nitrogen and sealed, then placed in boxes and vibrated on a transport simulator at 270 rpm for 14 min. This simulated truck transportation of about 1500 km. The structures tested were:

1. PE/Met PET/PE from various manufacturers.
2. PE/ EVOH/PE
3. Met PET/Met PET/PE

Structures 2 and 3 are considered as having high barrier properties. The second one has EVOH as a barrier and the third is a face-to-face metallized polyester. The latter structure is very stiff due to the face-to-face lamination of the metallized polyester. Permeabilities of the polymers tested are presented in Table I.

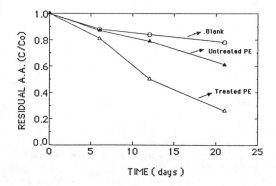

Figure 7.   The effect of untreated and oxidized polyethylene on ascorbic acid degradation in model solutions, stored at 35°C.

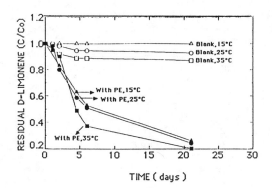

Figure 8.   The effect of polyethylene strips on residual D-limonene in a model solution, stored at different temperatures. (Reproduced with permission from Ref. 14.   Copyright 1987 Inst. of Food Technologists.)

Table I: Permeability values for different films

Film Manufacturer	Structure	Thickness (micron)	Permeability $(cc/m^2.24hr.At)$
A	Met PET/Met PET/PE	76	0.2
B	PE/Met PET/PE	105-110	1.9
B	PE/Met PET/PE	106-109	0.7
C	PE/Met PET/PE	115	1.0
D	PE/Met PET/PE	76	0.3
D	PE/Met PET/PE	94	0.5
E	PE/Barrier/PE	82	0.8
E	PE/Barrier/PE	75	<0.05

The juices, aseptically packed in the bag-in-box made from above films, were stored at 25°C.   Results of periodic tests made on these juices, as compared to juices packed aseptically in glass jars, are presented in Figures 9 -11. The juice in the structure Met PET/Met PET/PE deteriorated fastest as compared to the other structures.

The oxygen transmissibility increased significantly due to flex- cracking during transportation simulation, resulting in rapid ascorbic acid reduction. The transmissibility of these bags increased from 1.6 to 16.8 cc/day/bag after the vibration test. The juice in the other structures deteriorated at a slower rate but still much higher than in glass.   D-limonene in all films was reduced by 50% during the first 10 days at 25°C and after 25 days at 15°C and then remained constant.

These data conform with our previous results (14) as well as those from other investigators (7,8,11). The results show the combined effect of oxygen transmissibility and the interaction with polyethylene on a reduced shelf-life of citrus products, in aseptic packages made of various laminates, as compared to same products in glass.

Transport simulation.   The application of dynamic testing, which includes shock and vibration tests simulating distribution environments, is very important in evaluating package system-product performance.   In packages used for aseptic filling of citrus products, damages such as flex-cracking, pinholing, puncture and seal failure may occur. Any one of these events may result in an acceleration of the deteriorative reactions, or even leakage, resulting in total failure of the system.

For evaluating the effect of transport on the shelf-life of products in bag-in-the-box systems,two types of equipment were used:

1.   MTS 840 Vibration Test System. The bags were vibrated at 0.7 g and 10Hz for different times.

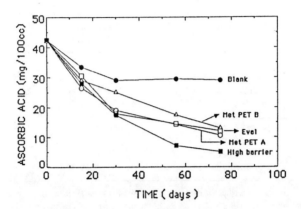

Figure 9.   Change of ascorbic acid in aseptically filled orange juice in bag-in-the-box after transportation simulation, stored at 25°C.

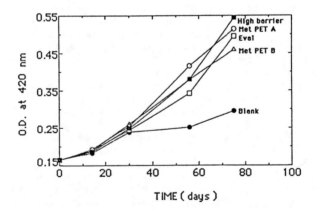

Figure 10.   Change of browning in aseptically filled orange juice in bag-in-the-box after transport simulation, stored at 25°C.

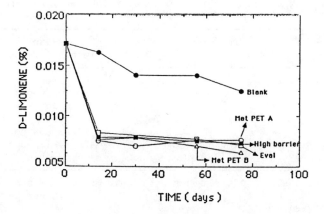

Figure 11 .Change of D-limonene content in aseptically filled orange juice in bag-in-the-box after transport simulation, stored at 25°C.

2. L.A.B. Vibration Transportation Simulator operated at a
   vertical-linear mode of 270 rpm, simulating travelling of a
   truck.
   In these tests we evaluated different structures of laminates
intended for bag-in-the-box with citric acid (5 and 15%), with
orange concentrate at 60°Bx and with water. Test conditions and
results are summarized in Table II.

Table II.   Transport simulation tests of "Bag-in-the Box"
            (Vibration conditions: 1) 0.7g,10Hz. 2) 270 rpm)

Structure	Product	Volume	Time (min)	Results
8%EVA/Met PET/PE (30/12/50)	15% cit.acid[1]	6gal.	90	Many pinholes, windows, leakage
	Water[1]	" "	2	Flex cracking, leakage
	Orange conc.[1]	" "	40	Flex cracking, leakage
PE/Met PET/8%EVA (30/12/50)	15% cit. acid[1]	6 gal.	90	Few pinholes upper part
	Water[1]	" "	40	Flex cracking
	Orange con.[1]	" "	40	" "
PE/Met PET/PE (50/12/50)	Water[1]	6 gal.	40	Flex cracking, pinhole
	Orange conc.[1]	" "	40	No damage
PE/Met PET/PE (50/12/50)	Water[2]	2.5 l	20	Flex cracking, pinholes, leakage
PET/Saran/Surlyn (50/12/50)	Water[2]	2.5 l	20	Pinholes, leakage
8%EVA/Met PET/8%EVA (50/12/50)	Water[2]	2.5 l	20	Flex cracking, pinholes, leakage
LLDPE/Met PET/LLDPE (50/12/50)	5% cit. acid[2]	2.5 l	10	Flex cracking, no leakage
PE/Met PET/PE (32/12/38)	5% cit.acid[2]	2.5 l	10	Flex cracking, no leakage

All the 6 gallon laminated bags, containing citric acid were
leaking after 90 min on the vibration test system. Also, large
"windows", demonstrated by disappearance of the aluminum layer,
appeared. Apparently these transport simulation conditions were
too severe. In another simulation test 6 gallon bags containing

orange concentrate were vibrated for 40 minutes at 0.7g and 10Hz.
These conditions simulated transport on a truck travelling over a
distance of 1000 km. All bags showed signs of flex-cracking and
three laminates had pinholes and were leaking. However, no
"windows" appeared in any of these bags. After testing same type
of bags on the simulator at 270 rpm for 20 min, simulating
travelling on a bumpy road for almost 2000 km, all bags leaked.
These again are very harsh conditions, and probably when testing
one should consider the distribution of acceleration amplitude (g)
and frequencies as created during transportation. After filling
the bags the flaps are folded causing flexing in the upper part of
the bag. In addition during vibration flex-cracking occurs
resulting in formation of pinholes in the internal PE liner and
also in the outer laminated bag. The acid medium creeps between
the layers, dissolves the aluminum creating the "window"
phenomenon in the bag. When the pinhole penetrates through the
outer layer, the bag starts to leak. This phenomenon was reduced
to some extent in bags containing viscous materials such as
concentrates, but when the medium was juice or citric acid, it was
very pronounced.

Summary

Shelf-life of aseptically filled juices in flexible and semi-rigid
plastic containers is limited by chemical reactions, which are
primarily influenced by storage temperature, as well as oxygen and
light. Absorption of oil soluble flavors by the contact layer and
accelerated deteriorative reactions caused by oxidized
polyethylene surfaces may also effect shelf-life. Results of
studies, relating to interactions of aseptically filled citrus
juices with various types of packages,are presented in this
paper. Reactions, such as ascorbic acid degradation and browning,
were accelerated in laminated cartons and in bag-in-the-box
systems as compared with glass jars. Also d-limonene
concentration was reduced to about half of its initial value
shortly after packaging in all laminates. Simulating
transportation, by dynamic testing, showed the damaging effect of
transport to the package resulting in reduced shelf-life.

Literature Cited
1. U.S. Food and Drug Administration. Fed. Register. 1981, 46, 6.
2. Schwartz, P.S. Food Technol. 1984, 38(12), 61.
3. Graumlich, T.R.; Marcy, J.E.; Adams, J.P. J. Agric. Food
    Chem. 1986, 34, 402.
4. Fox. R.W. In Aseptipack 85, Schotland Business Research Inc.
    Princeton. 1986, p 103.
5. Gilbert, S.G. Food Technol. 1985, 39 (12), 54.
6. Fernandes. M.H.; Gilbert, S.G.; Paik, S.W.; Steir F.
    J. Food Sci. 1986, 51, 722.
7. Durr, P.; Schobinger, U; Waldvogel, R. Lebensmittel
    Verpackung. 1981, 20, 91.

8. Marschall, M.R.; Adams, J.P.; Williams, J.W. In Aseptipak
    Schotland Business Research Inc. Princeton. 1985, p 299.
9. Buchner, N. Neue Verpackung. 1985, 4, 26.
10. Gherardi, S.T. Proc. Cong. of Fruit Juice Prod.. Munich,
    1982, p. 143.
11. Granzer, R. Proc. Cong. of Fruit Juice Prod. Munich,
    1982, 161.
12. Koszinovski, J; Piringer, O. Deutsche Lebensmittel
    Rundschau., 1983, 79, (6) 179.
13. Carlson, U.R. Food Technol. 1984, 38 (12), 47.
14. Mannheim, C.H.; Miltz, J.; Letzter, A. J. Food Sci.
    1987, 52, 737.
15. Passy, N.; Mannheim, C.H. J. Food Eng. 1983, 2, 19.

RECEIVED September 24, 1987

Chapter 7

Loss of Antioxidants
from High-Density Polyethylene

Its Effect on Oatmeal Cereal Oxidation

J. Miltz[1,2], P. Hoojjat[1], J. K. Han[1], J. R. Giacin[1], B. R. Harte[1], and I. J. Gray[1]

[1]School of Packaging, Michigan State University, East Lansing, MI 48824-1223

The loss of the antioxidants BHA and BHT from HDPE film was measured experimentally and analyzed theoretically. The volatilization of the antioxidant from the polymer surface was found to be the controlling parameter for mass transfer. Diffusion coefficients of the antioxidants evaluated from the experimental data and extrapolated to 100°C were found to be of the same order of magnitude as reported in the literature for BHT at that temperature. Oatmeal cereal packaged in a high level BHT impregnated HDPE had an extended shelf life when compared to a low level of BHT impregnated HDPE film, due to the adsorbtion, by the cereal, of the antioxidant that had evaporated from the package.

Antioxidants are widely used as food additives to retard oxidation of lipids and degradation of other components. Normally, the antioxidants are incorporated directly into the food. This has proven to be a very successful and inexpensive method for protecting oxygen sensitive foods.

Antioxidants are also incorporated into plastic films in order to protect them from degradation (1-2). It is well established that antioxidants are lost from polymeric films and sheets during storage (3-4). A relatively small amount of this antioxidant is lost through decomposition reactions (if not exposed to the outdoors environment) while the bulk of it is lost by what is commonly assumed to be a diffusion controlled process. The actual loss mode is, however, more complicated. For an antioxidant to be lost from a polymeric film (or sheet) it has to diffuse through the bulk of the polymer towards its surface and then

[2]Current address: Pillsbury Company, Research and Development Laboratories, Minneapolis, MN 55402; on leave from the Department of Food Engineering, Technion–Israel Institute of Technology, Haifa, Israel

evaporate from the surface into the surroundings. Depending on the nature and structure of the polymer and on the properties of the additive, to include its diffusivity and volatility, the loss process can be controlled either by diffusion or by volatilization or by a combination of the two.

The process of the physical loss of an additive which is soluble in the polymer thus involves two distinct processes: (i) the removal of additive from the surface by evaporation; and (ii) the replacement of additive in the surface layer by diffusion from the bulk polymer. A mathematical model describing the loss of an additive from the polymer to air requires therefore two parameters: a mass transfer coefficient characterizing transfer across the boundary of polymer surface-air interface and a parameter characterizing mass transfer within the polymer bulk phase.

Crank (3) has described a mathematical expression for a film from which additive is lost by surface evaporation with finite boundary conditions. According to this model, the total amount of additive leaving the polymer in time (t) is expressible as a fraction of the corresponding amount lost after infinite time by:

$$\frac{M_t}{M_\infty} = 1 - \sum_{n=1}^{\infty} \frac{2L^2 \exp(-\beta_n^2 T)}{\beta_n^2(\beta_n^2 + L^2 + L)} \tag{1}$$

where: M_t = amount of additive leaving the polymer in time = t
        M_∞ = amount of additive leaving at infinite time
        $T = Dt/\ell^2$ \hfill (2)
        $L = \ell\alpha/D$ \hfill (3)
        ℓ = half of film thickness
        t = time
        D = diffusion coefficient of additive in polymer
        α = volatilization mass transfer coefficient of additive from polymer surface

β_n values are the positive roots of the equation.

$$\beta_n \tan \beta_n = L \tag{4}$$

Calvert and Billingham (4) have analyzed the rate of loss of a simple low molecular weight moiety, such as the antioxidant 3,5-di-tertiary-butyl-4-hydroxy toluene (BHT), from thick slabs and/or bulk solids as well as from thin films and fibers. They were interested in polymer stabilization and assumed that polymer degradation will proceed rapidly to sample failure when the average concentration of additive falls to 10%, namely when $M_t/M_\infty = 0.9$. Ignoring terms other than n=1 in Equation 1, they have suggested the following failure criterion:

$$\frac{2L^2 \exp(-\beta^2 T)}{\beta^2(\beta^2 + L^2 + L)} = 0.1 \tag{5}$$

Calvert and Billingham (4) plotted the values of L as a function of T and concluded that at high values of L (thick film, rapid evaporation and low diffusion rate) the failure time is given by:

$$t = 0.87 \, \ell^2/D \qquad L>10 \qquad\qquad (6)$$

and is diffusion dominated and independent of α (or the rate of evaporation). At low values of L (thin films, slow evaporation and fast diffusion rates) they obtained a line represented by:

$$LogL + LogT = 0.383 \qquad\qquad (7)$$

leading to the failure time given by:

$$t = 2.42 \, \ell/\alpha \qquad L<0.6 \qquad\qquad (8)$$

According to Equation 8, the diffusion rate is unimportant at L values lower than 0.6 and the failure time is dominated by surface evaporation. They concluded that the rate of loss of low molecular compounds from a thick polymeric slab is determined by bulk phase diffusion, while the loss from thin films is dominated by the rate of surface evaporation.

The method of Calvert and Billingham (4) does not, however, enable one to determine the volatilization and diffusion coefficients separately as the parameters L and T in Equation 7 are interrelated. Han et al. (5) have extended the theoretical work of Calvert and Billingham and described a method to determine the two coefficients simultaneously from sorption or desorption data.

The present paper deals with the loss of the antioxidants 2-tertiary-butyl-4-methoxy phenol (BHA) and 2.6-di(t-butyl)-4 hydroxy toluene (BHT) antioxidants from a high density polyethylene (HDPE) film and the effect of the latter on the shelf life of oatmeal cereal packaged in flexible pouches fabricated from the test film.

Experimental

Materials. The HDPE films (density 0.959 g/cc thickness - 66 μm) were provided by Crown Zellerbach, Film Production Division (Greensburg, Indiana). The films contained 0.14% (W/W) BHA and 0.32% (high level) and 0.022% (low level) of BHT. The films with the two levels of BHT were used to prepare 18 x 19 cm pouches in which shelf life studies of the oatmeal cereal were carried out.

Oatmeal Cereal. Fresh oat flakes cereal were obtained from the Gerber Product Company (Freemont, Michigan).

Methods

Antioxidant Determination. The levels of BHA and BHT in the test film was determined by extraction followed by high pressure liquid chromatography (HPLC) analysis. The level of BHT was also determined by UV Spectrophotometry (Perkin Elmer Lambda 3B UV/visible with an integrating sphere) and by monitoring the change in film weight with time using a Cahn electrobalance.

For the antioxidant extraction, 5 g. of the film were cut into small pieces and extracted with 150 ml of acetonitrile in a Soxhlet extraction apparatus for 12 hours. The extracts were then filtered and diluted with the solvent to a constant volume of 200 ml. The

HPLC system consisted of a Perkin Elmer Series 3B Solvent Delivery
System and a LC-1000 Column Oven with a Perkin Elmer LC-85 Spectrophotometric Detector. The detector was interfaced to a Spectra
Physics/SP4200 Computing Integrator for quantitation. The chromatographic conditions were as follows:

Column a 0.24 x 25 cm ODS-HS sil-x-1 stainless steel (Perkin
Elmer)

Solvent 60% acetonitrile/40% distilled water (v/v)

Flow Rate 1 ml/minute

Detector Wavelength 291 nm for BHA and 280 nm for BHT

Injector 10 µl Hamilton Microliter #701-N Syringe

Peak areas and retention times were determined by the use of a computing integrator. The concentration of BHA and BHT in the film
samples was determined from standard graphs constructed by analyzing
pure BHA and BHT samples in acetonitrile.

Thiobarbituric Acid Analysis (TBA). The rate of lipid oxidation in
the cereal was measured according to the modified method of Caldwell
and Grogg (6). 20 g. of cereal were extracted overnight at room
temperature with 100 ml of hexane. The extracted lipid was filtered
using a glass suction funnel with an aspirator. The filtration was
then slowly evaporated in a rotoray evaporator maintained at 48°C.
The extracted sample was weighed and 100 ml of Benzene and 10 ml of
TBA reagent (prepared from 0.67 g. of TBA in 100 ml of distilled
water and 100 ml of glacial Acetic Acid) were added to it. The
sample was then vigorously shaken for four minutes and centrifuged
for 15 minutes; the top layer (benzene) was removed and the aqueous
layer transferred to a screw capped glass test tube and boiled for
30 minutes. The sample was then cooled before passing it through a
glass tube packed with cellulose powder that was used as a chromatographic column. Sample aliquots of 7.0 ml were forced through the
chromatographic column (by applying an air pressure of 10 psig) for
the separation of yellow and red components. The column was then
washed with 5 ml of distilled water to remove the yellow color. The
absorbed red fraction, associated with the oxidized components, was
thereafter eluted with 10 ml of aqueous pyridine (20%) and collected
in a volumetric flask. The optical density at 532 nm was read
against a blank and converted (calculated) to a 1 g. sample basis.

UV Spectrophotometer. A Perkin Elmer Lambda 3B UV/visible spectrophotometer with an integrating sphere was used to measure BHT
content of the films at different time intervals. Film samples were
mounted directly in the sample holder of the integrating sphere and
the absorbance (O.D. units) at 280 nm was recorded. The relative
concentration of BHT in the film was obtained by the expression

$$\text{Relative \% BH} = \frac{\text{O.D. (t)}}{\text{O.D. (o)}} *100$$

where the parameter in parenthesis is time.

Procedures

Studies of antioxidant depletion from the test films were carried

out on the BHA and high level BHT impregnated films. The film
samples were stored at different temperatures and the level of re-
tained antioxidant determined as a function of time. At different
times, film samples were either extracted for BHA or BHT analysis by
HPLC or analyzed directly by the UV spectrophotometric procedure.
Unless otherwise stated, the film samples were stored in open air.
For weight loss studies carried out with the electrobalance, the
samples hang inside a tube which was continuously purged with
nitrogen. The effect of BHT originally in the film on the rate of
cereal oxidation was carried out by analyzing the pouch and the
cereal for BHT content and extent of oxidation (TBA analysis) after
different storage period at 30 + 1°C and 45% RH.

Results and Discussion

In figures 1 and 2, the loss of BHA and BHT from the HDPE film is
shown as a function of time and temperature, respectively. The
straight lines in this semilogarithmic plots suggest that the loss
of these antioxidants follow a first order or pseudo-first order
rate expression:

$$\ln \frac{C}{C_O} = -Kt \tag{9}$$

where C_O and C are the initial and time T concentrations of antioxidant in the film samples (percent wt/wt) respectively; K is the
rate constant and t is the time interval. From figure 1 it can be
seen that nearly all of the BHA (greater than 95%) was lost within
one day at 50°C, within 3 days at 40°C and within 7 days at 30°C.
  The rate constants determined from Equation 1 for BHA are summarized in Table 1.

Table I.   Rate Constants for the Loss of BHA from HDPE
Film and the Activation Energy

Temperature (°C)	The Rate Loss Constant $K \times 10^3$ (1/hr)
10	4.0
22	9.6
30	19.6
40	46.6
50	121.8

To determine the BHT concentration in one of the HDPE films,
three different methods were used. These methods were: (1) BHT
extraction from the film followed by an HPLC analysis; (2) UV absorption of the film measured at 280 nm; and (3) direct measurement
of the film sample weight change with time. A linear relationship
between the results of the first two methods was found. In figure
3, the loss of BHT as a function of time as measured by HPLC and UV
is shown at two different representative temperatures. It can be
seen that both methods give comparable results. The loss of BHT
from the film, as measured by directly weighing the film on a Cahn
electrobalance, is shown in figure 4 for a third temperature.

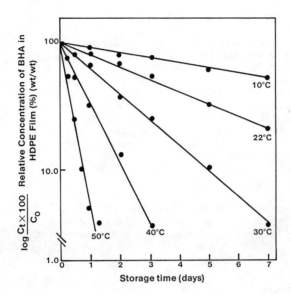

Figure 1. Loss of BHA from HDPE film during storage at different temperatures.

(Reproduced with permission from ref. 5. Copyright 1987 Society of Plastics Engineers.)

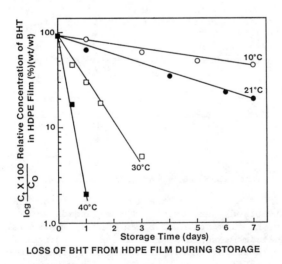

Figure 2. Loss of BHT from HDPE film during storage at different temperatures.

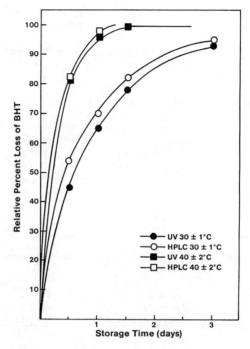

Figure 3. Loss of BHT from HDPE film as measured by HPLC and UV analysis.

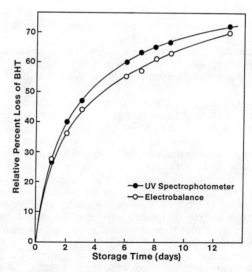

Figure 4. Loss of BHT from HDPE film: Comparison between UV and electrobalance results.

Comparable results to those obtained by UV spectroscopy were found
and thus all three methods can be used for BHT content determination.

The diffusion and volatilization mass transfer coefficients
determined according to the method outlined in a previous publication (5) are summarized in Table II for BHA and Table III for BHT.

Table II. Mass Transfer and Diffusion Coefficients for
BHA in HDPE Film

Temperature (°C)	Mass Transfer Coefficient αx10 (cm/sec)	Diffusion Coefficients Dx10 (cm^2/sec)
10	2.9	1.2
22	6.9	2.9
30	14.1	6.0
40	33.5	14.2
50	87.7	37.1

Table III. Mass Transfer and Diffusion Coefficients of
BHT in HDPE Film

Temperature (°C)	Mass Transfer Coefficient αx10 (cm/sec)	Diffusion Coefficients Dx10 (cm^2/sec)
10	3.9	2.2
21.5	8.6	4.8
30	36.4	20.0
40	152.6	80.0

(Reproduced with permission from ref. 5. Copyright 1987 Society of
Plastics Engineers.)

Figure 5 is a representative plot of the theoretical and
experimental rate loss curves for BHA and it can be seen that very
good agreement is obtained. Similar curves were obtained for BHT
at different temperatures.

No data for diffusion of BHA in HDPE were found in the literature.

Braun et al (7) measured the diffusion coefficient of BHA in
molten low density polyethylene (LDPE) by Inverse Gas Chromatography.
Although LDPE is different than HDPE and the activation energy for
diffusion in a molten polymer is different than that in the solid,
it was interesting to compare the calculated values for our experiments to those measured by Braun et al. It was expected that at
least the order of magnitude would be the same. It was assumed that
the melting point of LDPE is 108°C. Using an activation energy of
15.2 Kcal/mole up to 108°C and the value of 10 Kcal/mole reported by
Braun et al for the molten polymer, a diffusion coefficient of
3.5×10^{-7} cm^2/sec at 136.8°C was calculated. This value is 4.7
times higher than the values of 7.4×10^{-8} reported by Braun et al
for the same temperature. Taking into consideration our assumption
that BHA has the same diffusion coefficient in LDPE as in HDPE and
the same activation energy in the two polymers when in the solid
state, the agreement between the results is quite good.

Also for BHT, no diffusion coefficients in HDPE at room
temperature were found in the literature. Very recently Comyn et al

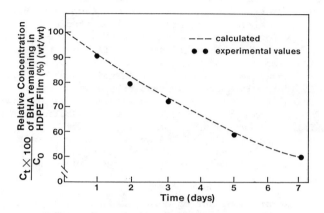

Figure 5. Comparison between theoretically calculated and experimentally determined results for loss of BHA from HDPE film.

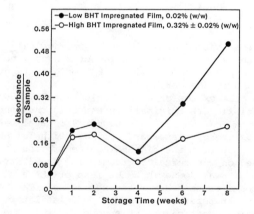

Figure 6. Oxidation results of oatmeal cereal packaged in high and low levels of impregnated BHT in HDPE film.

(8) reported a value of 12×10^{-8} cm^2/sec for the diffusion
coefficient of 2,6-di-t-butyl-4-methyl phenol (BHT) in HDPE at 100°C.
Although BHT and BHA are different additives, they are quite similar
in structure and are therefore assumed to have similar diffusion coefficients in HDPE or at least in the same order of magnitude. Using
the value of 15.2 Kcal/mole determined from our studies for the
activation energy for diffusion of BHA in HDPE, a value of 8.9×10^{-8}
cm^2/sec was calculated for the diffusion coefficient at 100°C. This
value is of the same order of magnitude as reported by Comyn et al
for BHT. Using an activation energy for BHT diffusion in HDPE of
12.6 Kcal/mole (determined from our results), a value of 10.3×10^{-8}
cm^2/sec was calculated, which is in good agreement with the results
of Comyn et al. No data for comparison of the volatilization mass
transfer coefficients were found in the literature.

To establish the validity of the evaporation/sorption mechanism,
storage studies were carried out with product containing tocopherol
and the product analyzed for BHT content. Table IV summarizes our
storage studies.

Table IV. Sorption of BHT by Cereal Product Packaged in
HDPE Pouches Following Storage at 39°C

Time (Weeks)	g BHT/g Cereal x 100	In Pouches	In Cereal	Lost to Environment
0	.0000	100	--	--
1	.0019	5	25	70
3	.0014	2	18	80
6	.0015	--	19	81

It can be seen that after one week of storage only 5% of the
BHT remained in the pouch. After 6 weeks of storage, no BHT could
be detected in the pouch but 19% of the BHT originally present in
the pouch was found in the cereal.

In Figure 6, the extent of cereal oxidation is shown for the
two levels of antioxidants originally present in the pouches. The
cereal contained in pouches with the higher BHT level was more
stable than its counterpart stored in pouches with low BHT content.
The BHT that was lost from the high BHT concentration pouches and
which had adsorbed on the cereal increased its stability.

To conclude, it was found that the loss of BHA antioxidants
from HDPE followed a first order rate expression. The controlling
parameter for mass transfer was found to be volatilization. The
calculated diffusion coefficients were compared to those reported
for BHT at 100°C and for BHA in molten LDPE and were found to be of
the same order of magnitude. The cereal contained in HDPE pouches
impregnated with the high level of BHT had lower oxidation levels as
a result of the migrated antioxidant from the package that had adsorbed on the cereal.

Literature Cited

1. Ram, A.; Meir, T.; Miltz, J. Intern. J. Polym. Mater. 1980, 8,
    323.

2. Waters Associates Inc. Oxidation of High Density Polyethylene by Weathering, bulletin No. 82276, 1981.
3. Crank, J. "The Mathematics of Diffusion". Claredon Press, Oxford, 1975, 2nd Ed.
4. Calvert, P. D.; Billingham, H. C. J. Appl. Polym. Sci. 1977, 24, 357.
5. Han, J. K.; Miltz, J.; Harte, B. R.; Giacin, J. R.; Gray, J. I. "Loss of 2-Tertiary-Butyl-4-Methoxy Phenol (BHA) From High Density Polyethylene Film". Polym. Eng. & Sci., In press.
6. Caldwell, E. F.; Grogg, B. Food Technology 1954, 185.
7. Braun, J. M.; Poos, S.; Guillet, J. E. Polym. Letters. 1976, 14, 257.
8. Comyn, J. Diffusion in Polymers, Plastics and Rubber Conference, 1986, 8-9 January, London.

RECEIVED October 27, 1987

Chapter 8

Overview of Sterilization Methods for Aseptic Packaging Materials

Romeo T. Toledo

Food Science and Technology Department, University of Georgia, Athens, GA 30602

Processes for effective inactivation of microorganisms on surfaces of aseptic packaging materials while operating in-line with the filling and package sealing equipment are described. Hydrogen peroxide is the only chemical sporicidal agent which is approved by regulatory agencies. Current commercial systems utilize 35% (w/w) hydrogen peroxide solutions and are effective. However, improper operation or design could result in spoilage or level of residues may exceed the tolerance. Ascorbic acid is a component most sensitive to residual hydrogen peroxide. Reduced hydrogen peroxide consumption and less problems with residues have been demonstrated in laboratory tests utilizing vapor phase hydrogen peroxide mixed with hot air as a sterilant. The synergistic effects of liquid hydrogen peroxide and high intensity ultraviolet radiation has also been demonstrated. Dry heat as hot air or superheated steam is suitable for sterilizing packages for low acid foods at $179.3^\circ C$ and for acid foods at $146^\circ C$ within the time frame suitable for high speed packaging systems. Moist heat at atmospheric pressure is effective on packages for acid foods when acidified water at pH 3.45 and $100^\circ C$ is used. Economic considerations and non-uniform dose delivery to pre-formed containers inhibit commercial adoption of ionizing radiation sterilization in-line with aseptic packaging systems.

Requirements for Sterilization and Regulations on use of Chemical Sterilants

Sterilization of packaging materials is a critical step in the aseptic processing and packaging operation. Although it is possible to pre-sterilize packaging materials using generally mild conditions for long times, this practice may expose the pre-sterilized packages to possible re-contamination when they are introduced into the aseptic packaging machine and therefore a second sterilization step is required just prior to use. Thus, in-line sterilization is preferred. Toledo (1) summarized the characteristics of sterilants suitable for in-line sterilization of packaging materials. The most important are: rapid microbicidal activity to enable high speed packaging; suitability for maintenance and control of parameters affecting microbicidal activity; compatibility with packaging material and equipment relative to corrosiveness and method for effective application; ease of removal from treated surfaces; no adverse effect on product quality; and no health hazards to workers around the packaging equipment or to consumers of food which may contain residues of the sterilant.

When using sterilants that do not leave any residue on the food contact surface, the Food and Drug Administration (FDA) consider sterilization as a process which is regulated only when used on low acid foods (2). However, when plastic packaging materials and chemical sterilants are used, the process is regulated as an indirect additive to food (3). The process of obtaining FDA approval for the use of chemical sterilants in food packaging has eased considerably since the concept was introduced in the early 70's. FDA's policy is still "show me" as expressed by Reister (4) and scientific evidence for process adequacy must be filed by processors before a system is placed into commercial operation. However, experience over the past 15 years has established the type of data that must be submitted and the procedures for obtaining these data have been well developed, therefore the process of obtaining FDA sanction for the process is not as cumbersome as it was before. As more data are developed on the spoilage patterns observed on commercially operating aseptic packaging systems, the criteria for sterilization will be more clearly defined and parameters for process adequacy can be effectively designed into a system. For example, traditional safety factors can be eliminated and target sterilization values can be reduced if uncertainties in the inter-relationships between the processing parameters and incidence of spoilage can be established. In addition, for aseptic packaging material sterilization, the traditional target sterilizing value equivalent to a "12 D process" should take into consideration the fact that good manufacturing practices in production and handling results in very low microbial contamination levels in packaging materials. The 12 D concept originated from earlier work (5) which established that multiplying by 12 the decimal reduction time (D) of <u>Clostridium botulinum</u> spores in dilute suspensions will be equivalent to the treatment time when no survivors can be recovered from 60 billion spores. Stumbo's (6) interpretation of "12 D" as a survival probability of 10^{-12} for a pathogen in a processed food is much too stringent. A widely accepted interpretation of the 12 D concept (7) involves setting an upper limit to the probability of spoilage from a pathogen at 10^{-9} per container. Thus an exposure time of 6D will be

equivalent to a 12D process in terms of product safety
(probability of survival is 10^{-9}) if the contamination level is
one spore in three containers (10^{-3}) in the packaging material.
     Hydrogen peroxide is now the only chemical sterilant for
packages that has proven to be acceptable to the consumer, the
processor, and the regulatory agencies in the US. When properly
used, it is an effective sterilant and level of residue can be
effectively controlled to within safe limits. FDA regulations (8)
limit the residual hydrogen peroxide to a maximum of 0.1 ug/mL
determined in distilled water assayed immediately after packaging
under production conditions. Initially, the use of hydrogen
peroxide as a sterilant for packaging material that directly
contact food was approved only for polyethylene (9). The approval
was extended to include all polyolefins in March 1984 (10), and in
1985 approval was extended to include polystyrene, modified
polystyrene, ionomeric resins, ethylene methyl acrylate copolymer
resin, ethylene vinyl acetate copolymer resin, and polyethylene
tetrapthalate (8). In January 1987 approval was extended to
include ethylene acrylic acid copolymers (11).
     Development of acceptable sterilization techniques suitable
for plastics, and public demand for more convenient good quality
canned foods, encouraged commercial adoption of aseptic packaging
technology. This review will discuss currently used practices for
in-line sterilization of aseptic packaging materials, critically
analyze newly developed procedures mentioned in the scientific and
patent literature that have potential, and discuss the influence
of the process used on product quality, storage stability and
properties of the packaging material.

Sterilization by Heat

Microbial inactivation in food preservation processes has
traditionally been accomplished by heating. Moist heat is
delivered by water or saturated steam. Since moist heat is
sporicidal only at temperatures above the boiling point of water
at atmospheric pressure, and since packaging machines are not
constructed to operate under pressure, it can only be used to
pasteurize packaging materials used for acid foods. Saturated
steam at atmospheric pressure and water close to the boiling point
may cause water vapor condensation which could interfere with
electronic components or with succeeding operations in the
formation of the package. Moist heat could cause blistering or
delamination of paper based packaging materials and impair the
heat sealing characteristics of plastics. Thus atmospheric steam
can only be used on non-paper based pre-formed containers. Hot
water is not used to pasteurize materials for packaging acid foods
because the resistance of non-sporeforming spoilage microorganisms
is maximum at pH values close to neutral (12).
     Data from our laboratory show that the microbicidal effect of
moist heat can be enhanced by the addition of organic acids to
water. Solutions of citric acid at a pH of 3.45 has been shown to

effectively eliminate cells of <u>Lactobacillus plantarum</u> and
<u>Leuconostoc mesenteroides</u> on filter paper strips after 90 s of
exposure at 100°C using an initial inoculum of 1 x 10⁷ cells.
<u>Saccharomyces cerevisiae</u> required only 15 s of exposure at 80°C
for complete inactivation. <u>Lactobacillus plantarum</u> cells on
polyethylene strips could not be recovered after 30 s exposure at
70°C but this result is probably due to cells being washed off the
strips rather than inactivation. Spore forming aciduric
microorganisms, <u>Clostridium butyricum</u> and <u>Bacillus coagulans</u> on
filter paper survived up to 90 s at 100°C when sub-cultured on
beef extract-tryptone-dextrose broth at pH 6.5 but the test strips
exposed to the same conditions but sub-cultured in apple juice (pH
3.2) did not cause spoilage. Although the above conditions
demonstrate the feasibility of citric acid solution to pasteurize
packages for acid foods, the necessity of operating at the boiling
point will require containment of vapors generated by the
sterilant solution in order that effective use can be made on an
aseptic packaging system. Sterilant residue will not be a problem
if a food grade acidulant is used.

Dry heat delivered by superheated steam or hot air is not as
microbicidal as moist heat at the same temperature, therefore
higher temperatures are required. When paper based packaging
materials are used, hot air is preferred as a sterilant over
superheated steam. The decimal reduction time for a heat
resistant yeast isolated from spoiled orange juice aseptically
packaged in spiral wound fiberboard container sterilized with hot
air is 0.038 min at 150°C and this time decreases by a factor of
10 for every 11.1°C increase in temperature (13). Thus, 12D for
this organism will require 27 s at 150°C or 1 min at 146°C. Dry
heat pasteurization of materials for aseptic packaging of acid
foods requires temperatures beyond the capability of most plastic
packaging materials to withstand.

When aseptically packaging low-acid foods, inactivation of
Clostridium botulinum is critical. In superheated steam, the
organism has a decimal reduction time of 1.2 min at 150°C and a
33.9°C increase in temperature is required to reduce the
inactivation time by a factor of 10 (14). An equivalent 12D
process based on an initial contamination level of 10⁻³ per
container and a probability of spoilage of 10⁻⁹ which could be
attained within a treatment time of 1 min will require a
temperature of 179.3°C. The exact values for inactivation of
various organisms in hot air and superheated steam may not be
exactly the same, but the magnitude is similar therefore
approximately the same conditions may be used when using either
forms of dry heat sterilization.

When sterilizing by hot air or superheated steam, there is no
effect on product quality. Furthermore, the temperature within
the entire aseptic filling chamber may be maintained at a level
lethal to microorganisms thus assuring asepticity. The high
temperature of the headspace gas at the time of sealing also
reduces the level of oxygen in the sealed package and can
contribute towards prolonging product shelf life.

Sporicidal properties of moist heat is much stronger than dry
heat, but cost of constructing a pressurized sterilization system

for aseptic packages is prohibitive. Yawger and Adams (15)
investigated the sporicidal properties of vapors generated from
aqueous solutions of glycerin and propylene glycol. They found
that although it is possible to generate vapor from these
solutions at temperatures higher than 120°C at atmospheric
pressure, the heat resistance of spores increases as the glycerin
or glycol concentration in the solution increases therefore the
effect of the high temperature is offset by the increased
resistance. For example, Yawger and Adams' data showed the
decimal reduction time of FS 1518 to be 3 min at 126.6°C when
exposed to vapors from boiling 86% propylene glycol solution while
the same organism required 0.84 min for one decimal reduction at
the same temperature in saturated steam.

Hydrogen Peroxide Solution

The sporicidal effectiveness of hydrogen peroxide solutions has
been well studied. Toledo, et al. (16) compared resistance of
spores of various microorganisms at both ambient and elevated
temperatures and reported the highest resistance by B. subtilis A
and B. subtilis var. niger. The latter forms yellow colonies
which eventually darkens as the culture ages, making it an ideal
organism for testing the capability of systems utilizing hydrogen
peroxide as a sterilant to achieve commercial sterility. Cerny
(17) reported a strain of B. stearothermophilus which was very
resistant to inactivation by hydrogen peroxide. This organism had
a decimal reduction time of 21.8 min in 30% hydrogen peroxide at
23°C compared to 5 min for a strain of B. subtilis. In
comparison, decimal reduction time in 35% hydrogen peroxide at
23°C calculated from our data (18) for B. subtilis A, is 6.4 min.
Since sterilization processes used on aseptic packaging systems
for low-acid foods must have documentation for adequacy to
inactivate pathogenic and spoilage microorganisms, a unit must be
tested using an acceptably resistant organism before it can be
placed in commercial production. The test organism must have a
resistance greater than spores of Clostridium botulinum in order
that the lethality of the process used for inactivation of the
inoculum would be equivalent to a 12 decimal reduction of C.
botulinum. For example, the decimal reduction time in 35%
hydrogen peroxide for C. botulinum 169 B, the most resistant of
the C. botulinum strains at 85°C is 0.034 min (19, 20) while that
for B. subtilis var. niger is 0.047 min (18), therefore 9 decimal
reductions of the latter organism is needed to give an equivalent
of 12 decimal reduction of the pathogen. Thus, the inability to
recover survivors in 100 inoculated test strips containing 10^{7}
spores of B. subtilis var. niger, would be evidence that the
process delivered a lethality equivalent to 12 decimal reductions
of C. botulinum type 169B.
     The exposure required to satisfy the above levels of
microbial inactivation to enable the packaging system to pass
microbial challenge tests for process validation, may vary
depending on how the hydrogen peroxide is applied and the
treatment which follows hydrogen peroxide application.
Inactivation of microorganisms occurs when adequate exposure time

is provided at the hydrogen peroxide concentration and temperature at the surface, prior to removal of the peroxide form the surface. Thus, although the exposure time is easily established in the laboratory when spores are directly added to hydrogen peroxide solution, the reality of not being able to directly monitor the presence of and concentration of hydrogen peroxide on a packaging material surface once the material leaves the dip tank prevents direct extrapolation of laboratory data to commercial practice.

In commercial aseptic packaging equipment which uses roll stock, the packaging material is immersed in hydrogen peroxide solution followed by heating to vaporize the peroxide before the packages are filled. One make of aseptic packaging machine is designed to apply the hydrogen peroxide solution by contact with a wetted roll. Contact time with the solution which contains a wetting agent, is often less than one minute. The wetting agent is added to the solution to enable the material surface to retain a film of hydrogen peroxide. As heat is applied to dry the surface the temperature of the hydrogen peroxide film is elevated promoting the microbicidal effect. Heat is applied using radiant electrical resistance heaters, a stream of hot sterile air, or by contact with a heated stainless steel roll. When acid foods are packaged the solution is maintained at ambient temperature. An elevated solution temperature may be used for low-acid foods. Cerny's (17) and Toledo et al's (16) data show that 1 min in 30% hydrogen peroxide at ambient temperature produced only 2.5 decimal reductions of mold spores and bacterial spores were not affected at all. Thus, in current commercial practice, the microbicidal action is primarily due to the heated film of hydrogen peroxide at the start of the drying phase of the sterilization process. For this reason, wetting of the packaging material and the presence of a uniform film of liquid on the material surface are critical factors.

Since there is no reliable system for directly monitoring the presence of a uniform fluid film on the surface of a packaging material, the rate of consumption of hydrogen peroxide is monitored instead. This method does not ensure that a uniform film is always present but it is based on the premise that if the system continually passed the microbial challenge tests for process validation on a measured rate of hydrogen peroxide consumption, then maintenance of that rate would be indicative of adequate levels of hydrogen peroxide on the packaging material to provide the sterilizing effect.

Hydrogen peroxide sterilized aseptic packaging systems which have been approved for commercial packaging of low-acid foods have been proven to produce an acceptable level of residues. However, occasionally inadequate drying results in residues exceeding the 0.1 ug/mL (8) tolerance and adversely affect product quality. An improperly designed system may result in more incidence of non-compliance with regulations pertaining to residues. Stefanovic and Dickerson (21) observed that removal of hydrogen peroxide from packaging material became more difficult as the temperature of the solution and immersion time increased. They postulated that more of the solution was adsorbed on the surface at high temperatures and the higher concentration of hydrogen peroxide vapor in air contacting the surface during the drying process prevented adequate removal of residues to meet the tolerance. Prolonged contact between liquid hydrogen peroxide and the packaging material also increased the difficulty of removing residues.

When sterilizing pre-formed containers, hydrogen peroxide is sprayed or atomized into the container. A measured amount of hydrogen peroxide is metered into each nozzle which delivers the solution into each container to ensure that a uniform film coats the inside surface of the package. Microbial lethality is induced during the drying phase of the process and the hydrogen peroxide vapors in the atmosphere surrounding the package could build-up to concentrations which could leave significant residues if the vapors generated are not displaced by dry sterile air.

Increasing the extent of microbial inactivation by increasing hydrogen peroxide concentration, temperature, or time of immersion may make residue reduction difficult to achieve in improperly designed systems. When reduced temperature and contact time in the hydrogen peroxide dip tank are used, consideration must be given to having the hydrogen peroxide solution remain on the surface for an adequate time such that at the surface temperature of the packaging material, commercial sterility can be achieved. For acid food products, the relatively low microbial lethality required makes it easy to achieve the desired level of microbial inactivation and meet the residue tolerance level. However, for low-acid foods, a system must be carefully designed such that both criteria of commercial sterility and adequate sterilant removal are achieved.

No data are available on the influence of hydrogen peroxide on the properties of the packaging material. Schumb, et al. (22) mentioned the suitability of polyethylene, polyvinylidene chloride, and polyvinyl chloride for storage of hydrogen peroxide at ambient temperatures, and the chlorinated polyethylenes at reasonably elevated temperatures. This indicates an inertness in these plastics for hydrogen peroxide. However these same authors also indicated a leaching of plasticizers from plastic by hydrogen peroxide. These indicate that although contact of a packaging material with hydrogen peroxide may not affect the polymer directly, it may affect the plasticizers and other additives, particularly at elevated temperatures, thereby influencing sealability and mechanical properties.

Residual hydrogen peroxide and those trapped in the package headspace at the time of sealing, has an adverse effect on product stability particularly on ascorbic acid degradation in fruit juices (23). Oxidative changes initiated by high levels of hydrogen peroxide may also affect other vitamins, essential oils and pigments, although no quantitative data is available for these effects. The concentration of residual hydrogen peroxide in the product determines the magnitude of these effects. When the concentration of hydrogen peroxide at the time of packaging exceeded 0.1 g/mL a marked increase in ascorbic acid degradation in bottled orange juice and orange juice concentrates was observed by Toledo (23). However, on some products where hydrogen peroxide is added directly to reduce microbial levels such as milk in cheese making and liquid whole eggs prior to freezing, storage stability is not affected, indicating that if hydrogen peroxide is

removed with catalase, only minor adverse effects may be observed.
Treatment of milk with 5000 g/mL of hydrogen peroxide at 49°C for
10 minutes lowered the cysteine and methionine content 10 to 25%
(24), but major vitamins except ascorbic acid were not affected.
50 g/mL hydrogen peroxide at 24°C for 8 hours was shown to have
no effect on vitamin content (except ascorbic acid) of milk (25).
Ascorbic acid was almost completely destroyed by addition of 3000
g/mL of hydrogen peroxide to milk and holding at 30°C for 24 h
(26). There have been no studies on storage stability of egg or
dairy products containing residual hydrogen peroxide.

Hydrogen Peroxide Vapor

It is difficult to maintain a record of all critical control
parameters during packaging material sterilization when the
process involves momentary contact with liquid followed by
heating. A large amount of hydrogen peroxide is also used during
the process and although in some systems the vapors are condensed
and eliminated through the liquid waste disposal system of the
processing plant, a substantial amount of the vapors escape to the
surroundings. Hydrogen peroxide vapor is a respiratory irritant
and environmental release should be curtailed. Although hydrogen
peroxide vapor release can be reduced by scrubbing or catalytic
treatment of the exhaust air, a process which uses the least
amount of hydrogen peroxide would be a cost effective alternative.
   Sterilization using hydrogen peroxide vapor mixed with air
would allow recycling if some means is employed to replenish
condensed vapors. Recycling would minimize environmental release
of hydrogen peroxide. Furthermore, the amount of hydrogen
peroxide adsorbed on treated surfaces from the vapor phase will be
several orders of magnitude smaller than a liquid film therefore
flushing the vapor treated surfaces with low temperature sterile
air free of hydrogen peroxide vapors can effectively eliminate
residues.
   Recently, a process was proposed where a mixture of hydrogen
peroxide solution and air was passed through a heated tube to
vaporize the hydrogen peroxide and the mixture was blown across
surfaces to be sterilized (27). This procedure generates a
mixture of hydrogen peroxide vapor and hot air which is used as
the sterilant. In this procedure however, it is difficult to
control the amount of hydrogen peroxide vapor in the air since a
high hydrogen peroxide vapor concentration facilitates excessive
condensation of liquid hydrogen peroxide when the mixture is
discharged into the packaging machine.
   Low levels of hydrogen peroxide which remain in the vapor
phase in air at near ambient temperatures has been shown to have
sporicidal activity suitable for in-line sterilization of aseptic
packaging materials (28, 29). Air saturated with hydrogen
peroxide vapor carry relatively low concentration of hydrogen
peroxide (7.6 mg/L at 70°C) but can induce 6 decimal reduction of
spores of B. subtilis var. niger in 1.2 min. In comparison the
same organism requires 1.2 min to achieve one decimal reduction in
hot air alone at 150°C. Wang and Toledo (28) bubbled air through
sintered glass immersed in hydrogen peroxide solution to generate
air saturated with hydrogen peroxide. This procedure ensures a

constant composition of hydrogen peroxide vapor in air and
conditions in the treatment chamber necessary to prevent
condensation can easily be determined since the condensation
temperature of the vapors in air is known. Since the amount of
hydrogen peroxide vaporized from solution is less than water, the
solution is enriched during prolonged operation and a system must
be devised to add make-up water to maintain the desired
concentration. Wang and Toledo (29) showed that a change in the
solution concentration from 35 to 37% increases the vapor phase
concentration from 7.6 to 8.2 mg/L and this change has an
insignificant effect on microbiological inactivation rate.

High-intensity Ultraviolet

It has long been recognized that ultraviolet light (UV) is
germicidal. UV irradiation has been suggested as a means of
decontaminating drinking water (30), maple syrup (31), apple cider
(32), and beef (33). Along with the development of high intensity
UV (34), data published by Bachmann (35) on dose levels required
to inactivate spores of molds and bacteria indicated excellent
potential of the process for sterilizing aseptic packaging
material. Maunder (36) suggested the use of high intensity UV for
aseptic packaging material sterilization and Cerny (37) reported
experimental data on the sporicidal properties of 254 nm rays from
a UV-C lamp. Using a source having an output of 30 mW/sq. cm.
yeast required only 0.5 s of exposure to achieve 5 decimal
reduction and 1 s induced 3 to 4 decimal reduction of spores of B.
stearothermophilus and B. subtilis. Conidia of A. niger were the
most resistant requiring 8 s to achieve 5 decimal reduction.
However, when very high level of inocula were used, a tailing
effect was found and first order inactivation rate was observed
only for the first 1 to 2 s of exposure. Survivors were recovered
from all the previously mentioned organisms after 16 min of
exposure. Shielding of organisms by those close to the surface
prevented inactivation. The same shielding effect might occur if
dust particles are present thus limiting the effectiveness of UV
irradiation for sterilization of aseptic packaging materials.
       The sporicidal properties of hydrogen peroxide is enhanced
when UV rays are applied on a surface wetted with hydrogen
peroxide solution. Cerny (37) showed that two more decimal
reduction of B. subtilis spores was achieved when packaging
material wet with 30% hydrogen peroxide at was UV irradiated
compared to UV treatment alone. The 5 s hydrogen peroxide dip
before UV treatment had no effect on spore count reduction.
Bayliss and Waites (38-40) also showed the same effect when
bacterial spores and non-spore forming bacteria in hydrogen
peroxide solution were exposed to UV irradiation. The process of
simultaneously applying hydrogen peroxide solution and UV will
provide adequate sporicidal activity to relatively low
concentrations of hydrogen peroxide in solution, minimizing the
problem of removal of residues after sterilization. No aseptic
packaging machine suitable for use on low acid foods which
utilizes the hydrogen peroxide-UV sterilization system is
presently in commercial operation in the US, although a US patent
has been issued to Peel and Waites on this process in 1981.

Ionizing Radiation

Gamma irradiation has been used to sterilize plastic bags for bulk packaging acid foods using an aseptic bag-in-box system (41). However, the difficulty of maintaining sterility during the transfer operation to the filling spout of an aseptic filling machine, raises some doubts as to the practicality of gamma irradiation sterilized packaging materials for use on low acid foods. The heavy shielding necessary to prevent radiation leakage from gamma ray sources makes the system impractical to use for in-line sterilization.

High energy electron beams are not as penetrating as gamma rays, and therefore does not require as much shielding as a gamma ray source. The microbicidal properties of ionizing radiation are the same regardless of source, and an effective sterilizing dose is in the order of 3 megarads. This dose level can be achieved with presently commercially available units (42) at rates suitable for in-line installation with a high speed packaging machine. However, with pre-formed containers, the geometry of the delivery system makes it difficult to apply a uniform dose on the whole container. A major impediment to adoption of this technology in the food industry is the high initial and operating cost of the units.

Care must be used when using ionizing radiation for sterilization of aseptic packaging materials. Although aluminum foil is not affected by irradiation, paper and plastic could be affected, with results ranging from discoloration to loss of pliability and mechanical strength. In general, polyethylene, ethylene co-polymers, and polystyrene are resistant to radiation damage at sterilization doses but flourine containing polymers, polyvinylidene chloride and polyacetals lose their tensile strength and become brittle and discolored (43)

Other Sterilants

There are a number of other compounds that have been suggested as havng potential for use in sterilizing aseptic packaging materials. Ethanol liquid and vapor has been suggested (44, 45). However, ethanol is effective only against vegetative cells and not against fungal conidia or bacterial spores therefore its use in packaging of foods is limited to extension of shelf life of packaged foods which are normally stored under refrigeration.

Peracetic acid is a compound produced by reacting concentrated hydrogen peroxide with acetic acid. On decomposition it forms acetic acid and water. The low pH and oxidizing properties make it an excellent sporicidal agent. Greenspan et al. (46) reported that a 50 g/mL aqueous solution of peracetic acid induced 5 decimal reductions of spores of $\underline{B}$. $\underline{thermoacidurans}$ within 1 min and 5 decimal reduction of $\underline{B}$. $\underline{stearothermophilus}$ within 5 min at 85°C. A 1% solution resulted in no recovery of spores of $\underline{B}$. $\underline{stearothermophilus}$ after 30 s exposure at 28°C. In spite of its sporicidal properties, peracetic acid is not an approved sterilant for use on aseptic packaging materials. Its vapor is very pungent and irritating therefore a system that utilizes this compound must be air-tight to prevent environmental release. Unlike hydrogen peroxide vapors for which a tolerance is allowed inside sealed packages, no such tolerance is allowed for peracetic acid and vapors in the headspace can cause a disagreeable vinegar-like off-flavor in some food products. In spite of these problems, this compound deserves attention as a possible sterilant for aseptic packaging materials because of its effectiveness as a sporicidal agent. The decomposition product has GRAS (Generally Regarded as Safe) status in foods, and in some foods, the presence of small amounts of acetic acid may not be considered an off-flavor.

Conclusions

Other methods for sterilization exist apart from the now FDA-approved heat sterilization and hydrogen peroxide sterilization techniques. However, these other processes require more costly systems to install and maintain, or regulatory approval is required to establish safety as an indirect food additive. Hydrogen peroxide is still the current sterilant of choice for plastic aseptic packaging materials and active research is continuing to devise ways of improving the effectiveness of the compound as a medium for achievement of commercial sterility in aseptic packaging systems.

Literature Cited

1. Toledo, R. T. Food Technol. 1975, 29 (5),102.
2. Code of Federal Regulations 1986a, Title 21, Part 113.
3. Code of Federal Regulations 1986b, Title 21, Part 174.
4. Reister, D. W. Food Technol. 1973, 27 (9), 56.
5. Perkins, W. E. In: Botulism, Proceedings of a Symposium; United States Public Health Service: Cincinnati, Ohio, Publication No. 999-FP-1, 1964; p 187.
6. Stumbo, C. R. Thermobacteriology in Food Processing; Academic: New York; 2nd ed. 1973; p 130.
7. Pflug, I.J.; Odlaug,T. E. Food Technol. 1978, 32 (6), 63.
8. Code of Federal Regulations 1986c, Title 21, Part 178.1005.
9. Federal Register 1981, 46 (6),2341.
10. Code of Federal Regulations 1984, Title 21, Part 178.1005.
11. Food Chemical News 1987, 28 (44), 28.
12. Hansen, N. H.; Riemann, H. J. Appl. Bacteriol. 1963, 26 (3), 314.
13. Scott, V. N.; Bernard, D. T. J. Food Sci. 1986, 50, 1754.
14. Stumbo, C. R. In: Industrial Microbiology; Miller, B. M., and Litskey, W., eds.; McGraw-Hill: New York, 1976; p 412.
15. Yawger, E. S., and Adams, H. W. Food Machinery Corp, Santa Clara, CA., Paper No. 179, 34th Annual Meeting, Institute of Food Technologists, New Orleans, LA, (1974).
16. Toledo, R. T., Escher, F. E., and Ayres, J. C. Appl. Microbiol. 1972, 26, 592.
17. Cerny, G. Verpackungs-Rundschau 1976, 27 (4), 27.
18. Toledo, R. T. In: Food Process Engineering; Schwartzberg, H. G., Lund, D. B., and Bomben, J. L. eds. ; AIChE Symposium Series, American Institute of Chemical Engineers: New York, 1982: 78 (218), 31.

19. Ito, K. A., Denny, C. B., Brown, C. K., Yao, M., and Seeger, M. L. Food Technol. 1973, 27 (11),58.
20. Denny, C. B., Personal communication. 1973. National Food Processors Association, Washington, D. C.
21. Stefanovic, S., and Dickerson, R. W. In: Current Technologies in flexible packaging ASTM STP 912, M. L. Troedel ed., American Society for Testing and Materials, Philadelphia, PA. 1986, 24.
22. Schumb, W. C., Satterfield, C. N., and Wentworth, R. L. Hydrogen Peroxide. Reinhold: New York, 1955
23. Toledo, R. T. J. Agr. Food Chem. 1986. 34, 405.
24. Tepley, L. J., Derse, P. H., and Price, W. V. J. Dairy Sci. 1958, 41, 593.
25. Gregory,M. E., Henry, K. M., Kon, S.K., Porter, J. W. G., Thompson, S. Y.,and Benjamin, M. I. W. J. Dairy Res. 1961, 28, 177.
26. Luck, H., and Schillinger, A. Z. Lebensm. Unters. Forsch. 1958, 108, 341.
27. Jagenberg, A. G. German Federal Republic Patent No. DE3235476C2, 1986.
28. Wang, J., and Toledo, R. T. In: Current Technologies in Flexible Packaging ASTM STP 912, American Society for Testing and Materials, Philadelphia, PA., 1986, 37.
29. Wang, J., and Toledo, R. T. Food Technol. 1986, 40 (12), 60.
30. Luckiesh, M., and Knowles, T. J. Bacteriol. 1948, 55,369.
31. Kissinger, J. C., and Willits, C. O. J. Milk and Food Technol. 1966, 29, 279.
32. Harrington, W. O., and Hill, C. H. Food Technol. 1968, 22, 1451.
33. Reagan, J. O., Smith, G. C., and Carpenter, Z. L. J. Food Sci. 1973, 38, 929.
34. Brandli, G. Brown Boveri Review 1975, 62 (5), 202 (Brown Boveri Corp., North Brunswick, NJ).
35. Bachmann, R. Brown Boveri Review 1975, 62 (5), 206.
36. Maunder, D. T. Food Technol. 1977, 31 (4), 36.
37. Cerny, G. Verpackungs-Rundschau, 1977, 28 (10),77.
38. Bayliss, C. E., and Waites, W. M. J. Appl. Bacteriol. 1979, 47,263.
39. Bayliss, C. E., and Waites, W. M. J. Appl. Bacteriol. 1980, 48,417.
40. Bayliss, C. E., and Waites, W. M. J. Food Technol. 1982, 17, 467.
41. Nelsn, P. E. Food Technol. 1984, 38 (3), 72.
42. Nablo, S. V., and Hipple, J. E. Technical Bulletin TN-21, 1972. Energy Sciences Inc., Woburn, MA.
43. Turianski, I. W. In: The Wiley Encyclopedia of Packaging Technology, M. Bakker, ed., J. Wiley, 1986, 562.
44. Doyen, L. Food Technol. 1973, 27 (9),49.
45. Mita, K., Japanese Patent 60256457, 1985.
46. Greenspan, F. P., Johnsen, M.A., and Trexler, P. C. Proceedings 42nd annual meeting of the Chemical Specialties Manufacturers Association 1955, 59.

RECEIVED December 9, 1987

Chapter 9

Theoretical and Computational Aspects of Migration of Package Components to Food

Shu-Sing Chang, Charles M. Guttman, Isaac C. Sanchez [1], and Leslie E. Smith

Polymers Division, Institute for Materials Science and Engineering, National Bureau of Standards, Gaithersburg, MD 20899

The numerical solutions and computational methods for the normal Fickian diffusion process applicable to packaging material is given in detail. Most experimental observations on the migration of small molecules from polymeric package materials into food or food simulating solvents show some non-Fickian behavior. In one case solvent absorption and swelling of the polymer have often been observed when the behavior is non-Fickian. A model for a solute diffusing in a swelling polymer is used to explain this phenomenon. In another case, where the migrant is sparingly soluble in the solvent, a stagnant solvent layer at the polymer surface may give rise to an initial migration behavior which is linear in time instead of linear in square root of time. In certain cases where the solvent is not stirred or is highly viscous, the quiescent migration is found to depend on the diffusion coefficient of the migrant in the solvent. Either alone or in combination, these models can be applied to describe most migration behavior in rubbery or semicrystalline packaging material.

The amount of package components that may be leached by food or food simulating solvents depends on the original concentration of the particular component or migrant in the polymer, its solubility in the solvent and/or the partition coefficient between the polymer and solvent as well as temperature and time. If the polymer is thick or the time is short, the amount migrated will be less than that predicted by the equilibrium partition. In these cases, dynamic modeling of the migration process is required to predict the migration as a function of time. In this paper we describe four

---

[1]Current address: Aluminum Company of America, Alcoa Center, PA 15069

models of the diffusion of migrant from food packaging material: the
simple Fickian diffusion with fixed boundary conditions, diffusion
in a material which is swelling, the effect of stagnant layer due to
low solubility of migrant into food or food simulant, and quiescent
migration into unstirred or viscous medium.

Solution for a normal Fickian diffusion process with simple
boundary conditions, as the migrant migrates from a plane sheet into
a stirred liquid, was solved long ago [1]. The numerical evaluation
of the solution generally depends on series expansions and converges
very slowly for early times or small amounts of material migrating.
To simplify the computational procedure, various special cases were
setup with limited ranges of applicability. With the ready
availability of personal computers, it is almost as convenient to
compute the general solution numerically without having to worry
about the limitations imposed by these special boundary conditions
and their ranges of applicability. A form of numerical procedure
for the solution of the Fickian diffusion applicable to a wide range
is shown in this paper.

For the diffusion of small migrant molecules from a polymer
into a solvent, or vice versa, there are often deviations from the
above described normal behavior. This is due largely to the
solvent/polymer interaction in the initial stage, which leads to a
swollen polymer. The initial stage is described by a smaller
diffusion coefficient which is a result of little or no solvent
content in the polymer. At later times the diffusion coefficient
increases to a higher value when the polymer is swollen by the
solvent in contact. A model is given which shows that the change in
the diffusion coefficient as a function of time follows closely the
movement of the solvent front in the polymer.

In the case of low solubility of the migrant in the solvent,
the rate of migration may be controlled by a thin stagnant layer of
solvent near the polymer surface. This stagnant layer generally
gives an initial migration behavior which is approximately linearly
proportional to the time, instead of to the square-root of time as
in normal Fickian diffusion behavior.

In some instances, where the solvent is not stirred or is
highly viscous, a quiescent migration phenomenon is found to depend
on the diffusion coefficient of the migrant in the solvent.

A combination of these cases is then possible to treat most
diffusion cases encountered in the additive migration behavior.

Non-interacting Systems-Normal Fickian Behavior

For a non-interacting system, the diffusion of a migrant between a
large plane sheet p of polymer of thickness L or 2ℓ and stirred
liquid s of finite volume V_s, the most widely used solution [1,2]
for the Fickian second law, $\partial C/\partial t = D\partial^2 C/\partial x^2$, is in the form of:

$$\frac{M_t}{M_\infty} = 1 - \sum_{n=1}^{\infty} \frac{2\alpha(1+\alpha)}{1+\alpha+\alpha^2 q_n^2} e^{-q_n^2 T} \qquad (1)$$

where $\alpha = M_s/M_p = V_s/KV_p$, the partition coefficient $K = C_p/C_s$ at $t \to \infty$
(assuming that the polymer and the liquid have the same density) and
the reduced time $T = Dt/\ell^2$. M, V and C denote the amount, volume

and concentration respectively. The concentration of the solute in the solution is assumed to be uniform. The concentration of the solute just within the sheet is K times that in the solution. The rate at which the solute leaves the sheet is always equal to that at which it enters the solution. The solution for the non-zero positive roots, q_n, of

$$\tan q_n = -\alpha q_n \qquad (2)$$

lies between $n\pi$ when $\alpha=0$ and $(n-1/2)\pi$ when $\alpha=\infty$. At $\alpha\ll1$,

$$q_n \sim n\pi/(1+\alpha) \qquad (3)$$

For other values of α,

$$q_n \sim [n - \alpha/2(1+\alpha)]\pi \qquad (4)$$

Computation of Equation (1) is quite straight forward with a computer. Equation (3) or (4) may be used to provide the starting value for q_n in a reiterative solution for Equation (2). At early times and small amounts of migration, the convergence of Equation (1) is rather slow, e.g., at $T \sim 0.001$ a sum of about 50 terms would be required.

Although the above mentioned calculation may not present any real problem in computation, there are however time savings from simpler solutions applicable for diffusion at early stages.

At $T<1$ and $\alpha<100$, the approximate solution [2]:

$$M_t/M_\infty = (1+\alpha) [1 - e^{T/\alpha^2} \operatorname{erfc}(T^{1/2}/\alpha)]$$

or its rational approximation for the error function [3]

$$M_t/(1+\alpha)M_\infty = 1 - \sum_{n=1}^{5} a_n \tau^n \qquad (5)$$

where $\tau=1/(1+0.3275911\ T^{1/2}/\alpha)$, $a_1=0.254829592$, $a_2=-0.28449636$, $a_3=1.421413741$, $a_4=-1.453152027$, $a_5=1.061405429$, may be used to converge quickly without creating significant error. However that this equation should not be applied at $T>1$ or $\alpha>100$.

As $\alpha\to\infty$, complete migration or migration into an infinite bath is expected, Equation (1) may be reduced to a function of T only, without the influence of partitioning

$$M_t/M_\infty = 1 - 2 \sum_{n=1}^{\infty} \frac{1}{q_n^2} e^{-q_n^2 T}$$

where $q_n = (n-1/2)\pi$. At $T<0.2$ or $M_t/M_\infty<0.5$, M_t/M_∞ is linearly related to $T^{1/2}$:

$$M_t/M_\infty = 2(T/\pi)^{1/2} \qquad (6)$$

For finite α or partition controlled migration, the range of linearity becomes much less.

Although these special cases have often been applied in the literature for the ease of reaching an estimate, Equation (1) should be used in most cases without worrying about the range of applicability. A combination of Equation (5) at T<1 and Equation (1) at T>1 gives most efficient computation for the entire region of normal Fickian diffusion process.

Variations of equations (1) and (5) often appear in literature concerning diffusion and migration. As $M_o = M_P + M_S$ and $M_\infty/M_o = \alpha/(1+\alpha)$, the amount migrated can be related to the total amount of migrant in presence instead of the equilibrium amount migrated at infinite time, thus $M_t/M_o = (M_t/M_\infty)[\alpha/(1+\alpha)]$. Amount migrated per unit area is another expression commonly used in migration studies, such that $M_t/A = LC_o (M_t/M_o)$. Concentration units are obtainable by the incorporation of the volume of the solvent and/or the polymer.

We have demonstrated that the Fickian behavior may strictly be followed [4], by eliminating the influence of the variance due to solvent/polymer interaction. After the polymer is first saturated with a radioactive liquid migrant, the diffusion of the labeled species are followed when the polymer is in contact with the unlabeled liquid migrant as the surrounding medium. In such a case, there is no net change in the composition of the solvent swollen polymer. Figure 1 shows the observed behavior of the migration of ^{14}C-labeled n-octadecane from branched or low-density polyethylene into unlabeled n-octadecane at 30°C, which follows strictly the behavior as described by Equation (1).

Solvent Swollen System

For the migration of small molecules from polymers, the above mentioned Fickian behavior is seldom obeyed, mostly due to solvent/polymer interactions. Smooth changes of diffusion coefficient with concentration of solution are often invoked to explain deviation from above described Fickian behavior. However the typical migration curve shown in Figure 2, for the migration of n-octadecane (1%) from polypropylene into ethanol at 30°C, is not explained by this effect. Two Fickian zones at times up to 0.2 h and after 10 h have been clearly observed. The diffusion coefficient of later times is enhanced due to the absorption of approximately 2% of the solvent by the polymer.

We shall try to understand the non-Fickian mass loss seen in Figure 2 from the polymers by studying the effect of mathematically coupling the solvent's swelling behavior to the diffusive behavior of the migrant. Rogers et al. [5] has studied the swelling of polyethylene films from solvents somewhat similar to those used by Chang and coworkers [6]. Rogers' work shows that the swelling of these films is distinctly non-Fickian. This behavior seems to be adequately characterized by the following two not mutually exclusive processes. After the semicrystalline polymer and the polymer-swelling solvent are brought together, there is a time interval during which Rogers finds that no measurable sorption takes place or during which the swellant is going in by normal Fickian diffusion. Following this induction period the mass uptake of the film by the swelling agent goes linearly in time, like case II diffusion [2].

In what follows we shall do some preliminary modeling to show that the coupling of these processes can adequately explain the experimental observations of Chang et al. Mathematical modeling which we will follow will be used to relate the experimentally

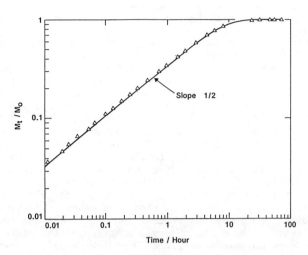

Figure 1. Migration of n-octadecane from branched polyethylene
into n-octadecane at 30°C.

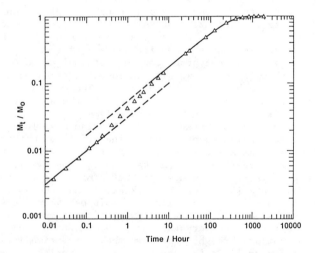

Figure 2. Migration of n-octadecane from polypropylene into
ethanol at 30 °C.

observed non-Fickian sorption behavior of the swelling solvent with the non-Fickian migration behavior of the additive. The sorption behavior of the swelling solvent will not be derived. Rather it will be taken as a known and the diffusion equation for loss of the additive will include the behavior of swelling solvent in the model.

From the above description the system has two time regimes. For early times the swelling agent goes into the polymer with normal case I diffusion. This time regime has been adequately described by the earlier work of McCrackin [7]. The result of that work suggests that there is an effective diffusion coefficient for the additive coming out. Thus for the time before the start up of the case II diffusion, t_0, we have the normal diffusion equation with an effective diffusion coefficient, D_2, as estimated from the McCrackin work

$$\partial C/\partial t = D_2 \partial^2 C/\partial x^2$$

with $C = 0$ at $x = 0$ for $t < t_0$.

For times greater than t_0, we propose a moving boundary problem which is defined as

$$\partial C/\partial t = D_1 \partial^2 C/\partial x^2$$

with $D = D_1$ at $x < S$ and $D = D_2$ at $x > S$, where D_2 is the effective diffusion coefficient, D_1 is the diffusion coefficient in the swollen medium, S is the position of the boundary between the two diffusion coefficients and the velocity, v, of the interface

$$S = v(t - t_0)$$

The diffusion coefficient in the unswollen region is assumed to be the effective McCrackin diffusion coefficient. Even without the additional complication of the McCrackin model the above equation with the moving boundary is extremely difficult to solve. Analytically few diffusion equations with a moving boundary conditions have been solved. Numerically the solutions of moving boundary diffusion equation show strong instabilities. We have been able to solve it in what we call an "adiabatic" approximation. In this solution the problem is solved as if the boundary at S were fixed in time and we obtain $M(t,S)$ at the boundary of interest. We then let $S = vt$ and then compute the $M(t,vt)$. These solutions are shown in Figures 3 and 4 for various diffusion coefficients and velocities of the front.

In Figure 3 the solution is given with no time lag allowed. It should be noted that the shape of curves is similar to those obtained experimentally by Chang et al. The change in effective diffusion coefficients seems to be correct but the range of data need to see both $t^{1/2}$ regions is much greater than seen experimentally. Also the rate of rise of the mass loss curve is much slower than seen in the experimental curve. Whether this difference from experiment is in the model or in the approximation to the model is difficult to say. We plan to explore this in a later paper.

Figure 4 show the effect on the same diffusion data of a 60

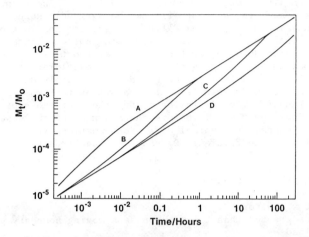

Figure 3. Mass loss in swollen systems. $D_1 = 16 \times 10^{-10}$ cm^2/s,
$D_2 = 10^{-10}$ cm^2/s
Time lag = 0 s
Solvent Front Velocity: A, 10^{-5}; B, 10^{-6}; C, 10^{-7};
D, 10^{-8} cm/s

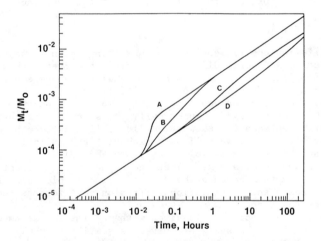

Figure 4. Mass loss in swollen systems. $D_1 = 16 \times 10^{-10}$ cm^2/s,
$D_2 = 10^{-10}$ cm^2/s
Time lag = 60 s
Solvent Front Velocity: A, 10^{-5}; B, 10^{-6}; C, 10^{-7};
D, 10^{-8} cm/s

second time lag. Clearly this sharpens up the data and gives a better shaped curve than seen in Figure 3.

We are currently exploring the numerical solutions to this problem and improvements on the model.

Boundary Layer Limited Migration

In migration experiments with solvent agitation, diffusion of the migrant in the polymer is usually assumed to be the rate determining step. This is not the only possibility. For example, if the solubility of the migrant in the food or food simulating solvent is limited, diffusion through a relatively stagnant boundary layer of solvent ℓ_s may become rate determining. The concentration profile for the boundary layer limited diffusion is as follows. The concentrations of the additive C_p and C_s in the bulk of the polymer and in the solvent are assumed to be constant irrespective to the distance from the interface and the stagnant layer at any time.

At the boundary, x=0, between polymer and solvent, we assume a partition equilibrium to exist, such that $C_p = KC_{s,0}$, as defined previously.

A necessary but not sufficient condition for the above concentration profile to prevail is for the difference in additive concentration between the bulks of polymer and solvent, $C_p - C_s$, to be very large compared to the concentration difference across the boundary layer, $C_p/K - C_s$. This condition prevails when K>>1 and we can approximate the concentration gradient across the boundary ℓ_s as a linear gradient.

By Fick's first law (flux is proportional to the concentration gradient), the rate at which the migrant migrates to the solvent is given by

$$dM_t/dt = 2AD_s(C_p/K - C_s)/\ell_s \tag{7}$$

where D_s is the diffusion coefficient of the migrant in the solvent boundary layer. The factor of 2 appears because diffusion occurs from both sides of the polymer sheet of thickness L.

By the use of the following relations:

$$C_p = (M_o - M_t)/V_p$$

$$C_s = M_t/V_s$$

and previously defined relations, now Eq. (7) can be rewritten as

$$\frac{d(M_t/M_\infty)}{dt} = \frac{2D_s}{K\ell\ell_s}\frac{M_o}{M_\infty}(1 - M_t/M_\infty)$$

which has the solution

$$M_t/M_\infty = 1 - e^{-\sigma t} \tag{8}$$

where

$$\sigma = \frac{2D_s}{K\ell\ell_s}\frac{M_o}{M_\infty} \tag{9}$$

At short times where $\sigma t \ll 1$, we have to a good approximation

$$M_t/M_\infty = \sigma t$$

or

$$M_t/M_o = 2(D_s/K\ell\ell_s)t$$

That is, M_t is initially linear in time.

A good procedure for determining σ in Equation (8) is to plot $\ln(1-M_t/M_\infty)$ against t. The slope of this plot is equal to $-\sigma$. In Figure 5, migration data for M_t/M_∞ values less than 0.9 of n-$C_{18}H_{38}$ from linear polyethylene into a 50/50 ethanol/water mixture at 60°C are plotted in this manner. From the slope, we obtain $\sigma=0.0002s^{-1}$. For this system we approximate that K = 1000, L = 0.07 cm, M_∞ = 0.18 and $D_s = 9\times10^{-6}cm^2/s$ and thus we find from Equation (9) that ℓ_s is about 70 μm. The diffusion coefficient D_s of n-octadecane in the 50/50 ethanol/water mixture was estimated from the Wilke-Chang correlation [8].

The idea of diffusion through a relatively stagnant boundary layer is not novel. For example, similar ideas have been invoked in theories of dissolution rates of crystals and electrode overvoltage [9]. Estimates of the thickness of the boundary layer for turbulent flow typically are in the 10 micron range [9].

To illustrate that Equation (8) cannot describe migration when diffusion in the polymer is rate determining, a plot of $\ln(1-M_t/M_\infty)$ against t for the migration of radiolabeled n-$C_{18}H_{38}$ from linear polyethylene into non-labeled n-$C_{18}H_{38}$ used as a solvent is shown in Figure 6. The strong deviation from linearity may be noticed for these data.

<u>Quiescent Migration</u>

We now address the situation where diffusion through the solvent phase is rate determining. This type of migration will be important in an unstirred solvent or when the food stuff is a solid or viscous liquid.

Consider an ideal system of an infinitely thick polymer in contact with an infinite solvent reservoir. There is only one polymer-solvent interface located at x=0. A partition equilibrium is assumed at the interface. On either side of the interface the flux is the same and is determined by the respective diffusion coefficient and the concentration gradient.

A general solution of this problem is

$$C_p = C_o \left[1 - \left(\frac{\mu}{\mu+1}\right)\left(1 - \mathrm{erf}\ \frac{x}{2\sqrt{Dt}}\right)\right]$$

or

$$C_s = C_o \left(\frac{\mu}{\mu+1}\right)\left(1 - \mathrm{erf}\ \frac{|x|}{2\sqrt{D_s t}}\right)$$

where

$$\mu = (D_s/D_p)^{1/2}/K.$$

By integrating the flux across the plane at x=0

$$J(0,t) = -\frac{\mu}{\mu+1}\ \frac{C_o}{\sqrt{\pi t}}$$

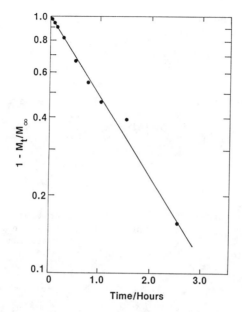

Figure 5. Migration n-octadecane from linear polyethylene into 50/50 ethanol/water mixture at 60 °C

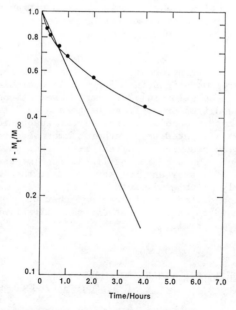

Figure 6. Migration of n-octadecane from linear polyethylene into n-octadecane at 30 °C

up to time t, the amount of additive, M_t, that will have crossed the plane at x=0 and of cross-sectional area A is

$$M_t = 4AC_o \ (\frac{\mu}{\mu+1})(\frac{Dt}{\pi})^{1/2}$$

The additional factor of 2 appears because we assume that both sides of the polymer sheet are in contact with the solvent. Although we assumed that the polymer is infinitely thick, all that is really required is for $L \gg \sqrt{Dt}$. Therefore, for sufficiently thick polymer sheets, since $C_o = M_o/AL$, we have

$$M_t/M_o = 2 \ (\frac{\mu}{\mu+1}) \ (\frac{T}{\pi})^{1/2} \tag{10}$$

For finite sheets for which $\ell < \sqrt{Dt}$ and/or finite solvent amounts, Equation (10) will slightly overestimate the amount of migration; thus, Equation (10) is always a conservative measure of quiescent migration.

Notice that at $\mu \ll 1$,

$$M_t/M_o = 2(D_s t/\pi \ell^2)^{1/2}/K.$$

The result for $\mu \gg 1$ is the familiar one for polymer limited migration with strong solvent agitation, as shown by Equation (6).

The above analysis of quiescent migration as well as boundary layer limited migration clearly shows that partition coefficients can have a strong influence on migration rates.

Conclusion

Most experiments on migration of small molecules for non-glassy polymeric packaging materials into food or food simulating solvents showed non-Fickian behavior [4]. In many observations, the behavior shifts from a low diffusion coefficient to a high diffusion coefficient due most probably to the solvent absorption by the polymer. We have simulated this rather common observations of a two-step process by assuming a solvent front that moves with a constant velocity. The behavior of linear in time can be explained by the existence of a stagnant solvent layer at the interface for cases of sparingly soluble migrants.

A combination of these processes and that of a quiescent migrant may be used to explain most cases of migration from non-glassy packaging materials.

Literature Cited

1. Crank, J. Phil. Mag. 1948, 39, 3621.
2. Crank, J. The Mathematics of Diffusion; 2nd Edition, Claredon Press: Oxford, 1975.
3. Abramowitz, M.; Stegun, I. A. Handbook of Mathematical Functions, Nat. Bur. Stand. Appli. Math. Ser. 55: Washington, D.C., 1972, p. 297.
4. Chang, S. S. Polymer 1984, 25, 209.

5. Rogers, C. E.; Stannett, V.; Szware, M. <u>J. Polym. Sci.</u> 1960, <u>45</u>, 61.
6. Chang, S. S.; Senich, G. A.; Smith, L. E. <u>Migration of Low Molecular Weight Additives in Polyolefins and Copolymers</u> (NBSIR 82-2472); NTIS PB 82-196403, National Tech. Info. Service: Springfield, VA 22161.
7. Smith, L. E.; Sanchez, I. C.; Chang, S. S.; McCrackin, F. L. <u>Models for the Migration of Paraffinic Additives in Polyethylene</u> (NBSIR 79-1598); NTIS PB 297671, National Tech. Info. Service: Springfield, VA 22161.
8. Wilke, C. R.; Chang, P. <u>AICHE J.</u> 1955, <u>1</u>, 264.
9. Vetter, K. J. <u>Electrochemical Kinetics</u>; Academic Press: New York, 1967.

RECEIVED November 5, 1987

Chapter 10

An Overview of Analytical Methods for Phthalate Esters in Foods

B. Denis Page

Health and Welfare Canada, Health Protection Branch, Food Directorate, Food and Research Division, Ottawa, Ontario K1A 0L2, Canada

The analytical determination of phthalate ester plasticizers as migrants in foods is reviewed. A typical procedure involves an extraction and column clean-up with quantitation by gas chromatography with electron capture or mass spectrometric detection. Distillation, gel permeation and other procedures have also been proposed to isolate the phthalate esters. Many investigators have noted the general contamination of laboratory supplies, solvents and reagents with phthalates, especially di(2-ethylhexyl) phthalate, to the extent that the detection limit for this phthalate in foods is determined by the reagent-laboratory blank. Methods of greatest sensitivity, therefore, should involve a minimum of reagents, simplified apparatus and minimal handling. Extraction, distillation and other isolation techniques are compared and the interference from pesticides and polychlorinated biphenyls is discussed.

The phthalate esters (PEs), the diesters of ortho-phthalic acid, are used almost exclusively as plasticizers, particularly for flexible polyvinylchloride (PVC). Plasticizers are substances incorporated in a material (usually a plastic or elastomer) to increase its flexibility, workability and stretchability. Generally, they are of low volatility and must be compatible with the polymers in which they are used. Some of the common PEs are listed in Table I. This listing is drawn from commercially available PE standards available in kits from various suppliers and reflects the PEs in general use.

PEs are conventionally prepared by the reaction of phthalic anhydride with a slight excess of a monohydric alcohol. A small quantity of sulfuric acid or other catalyst is necessary for complete formation of the diester. Some PEs are produced using a mixture of branched chain isomeric alcohols as indicated by (M) in

0097–6156/88/0365–0118$06.00/0
Published 1988 American Chemical Society

Table I.   Common Phthalate Esters and Their Abbreviations

Structure	Ester	Abbreviation
	dimethyl	DMP
	diethyl	DEP
	di(n-butyl)	DNBP
	diisobutyl	DIBP
	diallyl	DAP
	diisohexyl (M)	DIHP
	dicyclohexyl	DCHP
	diphenyl	DPP
	di(n-octyl)	DNOP
	diisooctyl (M)	DIOP
	di(2-ethylhexyl)	DEHP
	diisononyl	DINP
	diisodecyl (M)	DIDP
	diundecyl	DUP
	ditridecyl	DTDP
	butyl benzyl	BBP
	butyl octyl	BOP
	butyl isooctyl (M)	BIOP

The structure shown at left is a benzene ring bearing two ester groups: CO_2R and CO_2R'.

Table I.   Furthermore, PEs such as butyl benzyl phthalate, can be produced in which the two ester moieties are different.

In 1985 the consumption of PEs in the U.S.A. surpassed 500,000 metric tons (1).   Of the PEs, DEHP is the most widely used, although its share of the total phthalate market has been declining.   DEHP and related isomeric octyl phthalates account for over 50% of the phthalate plasticizers produced.

DEHP, the most important PE is a demonstrated carcinogen in rats and mice (2).   The implications to humans, however, have yet to be determined.

DEHP, DEP and DIOP have been granted prior sanction as plasticizers in the manufacture of food-packaging materials (for foods of high water content only) and are listed in Sec. 181.27 of the U.S. Code of Federal Regulations (CFR).   Other phthalate esters have been cleared as plasticizers under Sec. 178.3740 of the CFR for various food-contact uses.   In Canada, such plasticizers are regulated under Division 23 of the Food and Drug Regulations. DEHP-plasticized PVC, for example, has been sanctioned by the Canadian Health Protection Branch for use in certain aqueous food contact applications by the issuance of no-objection letters respecting its use.   Some of the permitted food uses for DEHP-plasticized PVC in the U.S. are produce wrap; bottle cap or lid liners for beer, soft drinks, juices, baby food and infant formula; and tubing for milk transport or carbonated beverage contact.   The use of DEHP as a plasticizer in food-use applications, however, has been declining and it is now little used.

Diethylhexyl adipate (DEHA), a plasticizer not based on phthalic acid, has also been cleared under Sec. 178.3740 of the CFR and is currently the most widely used plasticizer for food

packaging. DEHA is used principally as a plasticizer in PVC- and
polyvinylidene chloride-based food wrap.   Such wrap is used for
retailing fresh meats, poultry and fish.   The use patterns of
plasticized food-contacting materials are constantly changing as
new materials and processes are developed.
    In summary, analytical procedures for the phthalate esters and
DEHP in particular have become important for a number of reasons.
Firstly, DEHP is a potential health hazard to man. Secondly, DEHP
and other phthalate esters are produced worldwide in large
quantities and used extensively as plasticizers.   Finally, as a
result of its toxicity and large annual production, DEHP has become
an environmental concern.
    Analytical procedures are required by regulatory agencies who
are responsible for ensuring public safety by monitoring foods for
excessive and potentially harmful levels of PEs.   These esters may
be actual background levels from environmental sources or contributed to the food from food-contacting plastics during the production, handling, processing, transporting or packaging etc. of the
food.   Furthermore, methodology is required to establish data bases
to evaluate changing residue levels and use patterns as well as to
calculate dietary intakes.
    Analytical methods for the analysis of PEs are the key in
studying the migration of packaging components from the package or
food-contacting material into the food.   Questions, such as the
following can only be answered using the appropriate analytical
methodology: What is the level of migrating component in the food?
How rapid is the migration from the package to the food?   Is the
migrating packaging component distributed homogeneously throughout
the food or is it concentrated near the surface? By analyzing the
package we can identify and quantitate possible migrating packaging
components.   Decreases in the level of the component of interest
resulting from food contact can also be monitored.   From the above
discussion, aside from any regulatory aspects, the importance of
analytical methods in studying the interaction of packaging components and foods is readily apparent.

Analytical Procedures

Procedures for the analysis of PEs in foods invariably include
DEHP, and, as described below, are derived from multiresidue
pesticide procedures.   Procedures have been developed for other
esters or plasticizers, but these are usually specific studies,
concerning packaging or toxicology, in which the analytical method
is used to study migration (packaging) or level, distribution or
excretion vs. time (toxicology).   Such procedures are an important
source of methodology which may be applied to foods.   Similarly,
environmental analytical procedures for PEs, in which various
compounds are determined in air, sediment or water also provide
analytical information.   In these situations, PEs are the target
analyte, whereas in pesticide procedures they are of secondary
importance.
    DEHP in particular has become a widespread laboratory con-
taminant due to its extensive use as a plasticizer for non-food
contacting PVC.   This was recognized in the 1960's and became the

subject of several reports in the 1970's and 80's. Giam (3) reported in 1975:

> "This problem of background contamination has been more serious in the trace analysis of phthalates than in studies of many other pollutants (including the chlorinated hydrocarbons) because phthalates are present in almost all equipment and reagents used in the laboratory".

His study concerned the low ppb analysis of DEHP in open-sea fish. The occurrences of DEHP, DNBP and other PEs in a variety of common laboratory reagents and supplies have been reported by many workers. Table II summarizes the findings of two of these reports (3,4). There is some agreement between the findings of the two investigations, however, one must consider the contamination to be variable. It is important to note that PEs may be components of some of these materials, e.g. plasticized tubing, and not

Table II. DEHP and DNBP in Common Laboratory
Materials and Reagents

Material	DEHP	DNBP
	percent	
Tubing - elicon	67	
- PVC	11.6,16.7	23.3,10.4
- black rubber	0.3	
	$\mu g/g$	
Tubing - latex	4	
- polyethylene	0.8	
Cork	6	
Glass wool	1,4	0.8
Filter paper	0.6	5
Extraction thimble	1.3	500
Chromatography paper	2.3	100
Tissue		42
Aluminum foil	0.3-4.9	0.1-1.6
Screw cap	0.7	25
Teflon sheet	0.4	
	ng/g	
Silicic acid	1000,2400	200,600
Alumina	45	42
Florisil	0.05,64,88	24,92
Sodium sulfate	2-40	24-90
Sodium chloride	1.5-110	
Tap water	1.5,3.8	1.9

contaminants in the true sense of the word. In chemical reagents
the PEs are considered contaminants. DNBP is the most volatile PE
used as a plasticizer (5). The more volatile DMP and DEP are not
used. DNBP is reported to be readily lost by volatilization from
plasticized PVC, probably accounting for its occurrence as a common
laboratory contaminant. Giam (3) reported that airborne
contaminants could readily recontaminate previously cleaned glassware. DEHP was specifically mentioned.

PEs are also present as components in a wide variety of common
items found in the laboratory such as instrument cables, disposable
plastic gloves, instrument covers and plasticized caps for
laboratory supplies. Items such as these may be responsible for
the ubiquitous presence of DEHP in the laboratory environment. The
presence of PEs, especially DEHP and to some degree DNBP, and their
associated procedural blank levels are undoubtedly the most proble-
matic aspect of trace PE analysis. These procedural blanks reflect
the PE contribution of all the reagents, materials and equipment
used in the determination as well as that of the laboratory
environment. Procedural blanks must be subtracted from the
analytical result to give the actual level of PE. It is important
that these procedural blanks be as low and as consistent as
possible and they must be run simultaneously with the analytical
determination. For analysts interested in background or environmental levels of DEHP, this problem of background contamination is
formidable; yet for those determining the more uncommon plasticizers, background contamination is not a major problem. The study
of migration from the package to the food also requires careful
procedural controls, especially if DEHP or other common PEs are to
be studied. In this case a procedural blank, a food blank, and the
analysis of the contacted food are required. If isotopically
labelled DEHP, indeed an uncommon PE, can be incorporated into the
packaging material, then, with suitable detection, the background
problem is minimized.

Giam has described a comprehensive protocol to minimize
procedural blanks (3). Organic solvents and rinsings of solid
reagents were monitored for purity after concentration using gas
chromatography with electron capture (GC-EC) detection. Pesticide-
residue-free grade of organic solvents were found to be
sufficiently pure for use. Diethyl ether, however, required
distillation just before use. Water was purified by extraction.
Inorganic salts and Florisil were heated and stored at 320° before
use. Glassware, because of its large surface area, was considered
to be a major source of contamination. Glassware, as well as other
equipment, required washing, an aqueous rinse, and a 4-fold acetone
rinse. Teflon parts were washed, rinsed and wrapped in clean
aluminum foil. All other equipment and aluminum foil were heated
at 320°C for 10 h before use. Glassware and equipment were covered
with clean aluminum foil while cooling. The cleaning efficiency
was monitored by rinsing the equipment or glassware with petroleum
ether. The petroleum ether was then concentrated to monitor
residual contamination.

The variety of PEs available as standards in Table I reflects
those determined in foods or environmental samples as described in
the published analytical literature. Reports including DEHP and

DNBP in the analytical scheme are most common. These phthalates are also included in the general analytical schemes of the Canadian Health Protection Branch (HPB) Analytical Methods for Pesticide Residues in Foods (6). The U.S. Food and Drug Administration (FDA) Pesticide Analytical Manual (PAM) has included 9 phthalate esters (DMP, DEP, DIBP, DNBP, BBP, DEHP, DNOP, DIHP and DIOP) in their multi-residue procedure (7). This is probably the most complete listing of phthalate analytical data and includes relative retention times on several different columns, electron capture detector sensitivity, as well as recoveries using selected multi-residue procedures. The Chemical Manufacturers Association is currently studying the environmental effects of 14 commercially important PEs (8). They have concluded that all 14 esters, DMP, DEP, DNBP, BBP, DNHP (di-n-hexyl phthalate), BOP (butyl 2-ethylhexyl phthalate), di(n-hexyl, n-octyl and/or n-decyl) phthalate, DEHP, DIOP, DINP, di(heptyl, nonyl and/or undecyl) phthalate, DIDP, DUP and DTDP, exhibited a low potential for adverse effects on the environment, and a low to moderate potential for bioaccumulation.

The above listings of PEs demonstrates the wide variety of these chemicals that are of concern to different researchers, regulatory agencies or associations. These PEs, as a group of compounds, have a wide range in volatility and polarity such that specific analytical procedures applicable to PEs with similar physical properties may not be suitable for others. However, consideration of the differences, should permit modification of methods to suit the phthalate ester(s) of interest.

Analytical Methodology

Methodology for pesticide residues often includes gas chromatographic retention data, detector sensitivity and recovery data for the more common PEs, probably because of their occurrence as a laboratory contaminant rather than a residue. Consequently, when methodology is required for PEs, the typical proven pesticide-type procedures are often chosen.

Pesticide methods as applied to PEs in fatty foods in the above governmental publications typically involve sample preparation, a lipid extraction step, a liquid-liquid partition and/or a column cleanup step, and a final determinative step using GC-EC detection. Different sampling approaches are required depending on the physical state of the sample and on what residue information is required. Appropriate homogenization of the sample may be necessary if the analyte concentration, e.g. in $\mu g/g$, is to be determined on the whole product. This is especially important for solid or semisolid foods in which packaging migrants may be found concentrated near the food surface. To study the migration of a packaging component into solid or semisolid food, a core sample of the food in a large package or a food sample packed in a non-plastic package would provide an appropriate blank. The blank could then be compared to a surface layer sample and the results expressed in mg/dm^2 of migrating packaging component. The extraction step will depend upon the food analyzed. In recent years concerns over the migration of DEHP from food-contacting plasticized resins, e.g. milk transport tubing, to milk or milk products has required appropriate analytical procedures to monitor DEHP in milk or cheese. Similarly, DEHP in plasticized cap liners may also migrate into contacted foods. Figure 1 outlines typical pesticide-type procedures which may be used to determine DEHP and other PEs in cheese (fatty food) and beer (nonfatty food).

In Figure 1A, which is abstracted from the HPB publication (6), 100 g of a representative and/or homogeneous cheese sample is extracted by blending with 300 mL of methylene chloride and 50 g of anhydrous sodium sulfate. The supernatant is then filtered and evaporated to obtain the solvent-free fat. Alternative procedures are available to isolate fat from cheese. For example, in the FDA PAM procedure 211.13c, the sample is blended with sodium or potassium oxalate and alcohol and then extracted three times with ethyl ether and petroleum ether. The combined extracts are washed with water, dried through sodium sulfate and evaporated to obtain the fat. This extracted lipid material also contains any co-extracted PEs, pesticide residues, and environmental contaminants such as the polychlorinated biphenyls (PCBs), and must be subjected to further fractionation to separate the non-polar lipids from the compounds of interest.

In the HPB procedure, two fractionation procedures are described. In one procedure, 1 g of extracted lipid material in hexane is directly applied to a 2% water-deactivated Florisil column. Sequential elution with hexane, increasing percentages of dichloromethane in hexane, increasing percentages of ethyl acetate in hexane, and finally ethyl acetate are used to separate the phthalate esters from the majority of the pesticides, the PCBs as well as sample coextractives. In order to process an initial fat sample greater than 1 g on the Florisil column described above, a preliminary acetonitrile (AN)-hexane partition is necessary to reduce the lipid content of the sample. Thus, 3-4 g of the isolated fat in 40 mL of hexane is extracted with 4 x 40 mL of AN. To the combined extracts in a separatory funnel, dichloromethane, water, a phosphate buffer and hexane are added, the funnel shaken, the aqueous layer removed and the organic phase rinsed with water. This extract is concentrated under vacuum and diluted with hexane before Florisil cleanup as described above. The equivalent FDA procedure is described in 211.14a (petroleum ether-acetonitrile partition) followed by Florisil chromatography (211.14d).

The Florisil column will separate the phthalate esters from the PCBs and the majority of the pesticides. Some fractionation of the short-chain PEs from the longer chain esters will occur. In the HPB procedure 7.1(a), DNBP and DEHP are eluted with 5% ethyl acetate in hexane and in the PAM procedure 211.14d the PEs are typically found in the 15 and 50% ethyl ether in petroleum ether eluants. However, some of the DEP is reportedly present in the 6% ethyl ether fraction. Gel permeation chromatography (GPC) is a second procedure that can be used to separate lipid material from pesticides and PCBs. However, neither government publication includes data for the PEs. After GPC further fractionation on a Florisil column would be necessary to separate the PEs from the pesticides, PCB's etc. As the determinative step, i.e. GC-EC or gas chromatography with mass spectrometric (GC-MS) detection, is also applicable to extracts of non-fatty foods such as beer, this final step will be considered after a discussion of Figure 1B.

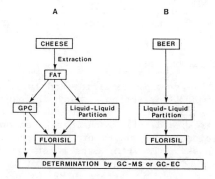

Figure 1.   Example of Pesticide Procedures Applied to Phthalate Ester Determinations in A, Fatty Foods and B, Non-Fatty Foods.

Figure 1B describes a typical procedure (PAM 212.13a) for non-fatty foods and is applicable to the determination of DEHP, and other possible PE contaminants in beer. Beer contains no solid particles and requires only decarbonation, partitioning and Florisil chromatography. 50 g of beer and 200 mL acetonitrile are blended and transferred to a separatory funnel. 100 mL of petroleum ether is added and shaken with the acetonitrile-beer solution. 10 mL of saturated aqueous sodium chloride and 600 mL of water are added and, after mixing, the aqueous layer is discarded. The organic phase is washed twice with 100 mL of water and dried over sodium sulfate before Florisil column chromatography (PAM 211.14d) and GC-EC or GC-MS.

In all the above procedures, care must be taken to avoid loss of the PEs, especially those of greater volatility, i.e. lower molecular weight. Spiked and unspiked food matrix blanks are employed to ensure adequate recoveries for each particular food and PE.

Chromatography and Detection

The determinative step in pesticide and in PE analysis usually involves GC-EC or GC-MS detection. Many different liquid phases have been employed to separate the various PEs and considerable information is available detailing their separation and that of the coextracted pesticides on a number of different packed column gas chromatographic phases. The PAM lists relative retention data on three different packed column liquid phases for 10 phthalate esters including the mixed isomeric esters DIHP (4 peaks) and DIOP (7 peaks). The three liquid phases, OV-101, OV-17, and OV-225, represent non-polar (methyl silicone), moderately polar (50% methyl 50% phenyl silicone) and fairly highly polar (25% cyanopropyl 25% phenyl 50% methyl silicone) columns, respectively. Since about 1980, when the first fused silica capillary columns were introduced, capillary GC has become widely used for PE and pesticide analysis. The chromatograms in Figure 2 were obtained in the author's laboratory using bonded capillary phases essentially equivalent to the OV phases listed above. Nine PEs are detected using EC detection and separated in temperature programmed runs from 80°C (2 min hold) to 300°C, 260°C and 240°C at 10°C/min for the DB-5 (1 μm film), DB-17 (0.25 μm film) and DB-225 (0.25 μm film) phases, respectively. Each column was 15 m x 0.32 mm i.d. The separation in Figure 2A is carried out on the DB-5 bonded phase which is a 5% phenyl methyl silicone and is considered very slightly polar. On this column the PEs generally elute in the order of boiling point or molecular weight, the more highly branched isomers eluting before those with straight chains. On the DB-5 column, BBP elutes well before DEHP, and DPP elutes well before DNOP. On the moderately polar DB-17 column (Figure 2B), BBP now elutes just before DEHP and DPP elutes after DNOP. On the more polar DB-225 column (Figure 2C), BBP now elutes well after DEHP, and DPP well after DNOP. Thus, as the phenyl content of the bonded phase increases, the relative retention of BBP and DPP, the two PEs with aromatic ester groups, also increases. This demonstrates the

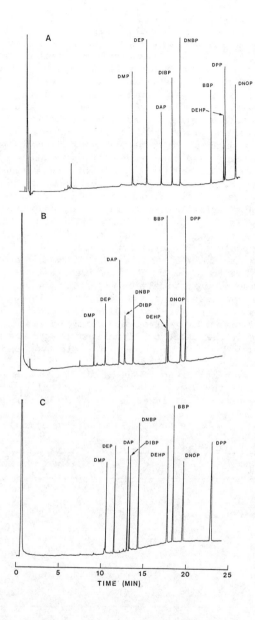

Figure 2.   Gas Chromatograms of Phthalate Esters on   A, DB-5;
B, DB-17; and C, DB-225 Capillary Columns.   See text for
details.

usefulness of using columns of different polarity for confirmation
purposes as well as for possible separations of PEs from food
matrix interferences. No more than a dozen capillary column bonded
phases are available for phthalate or pesticide applications
compared to the over one hundred liquid phases for packed column
use. The phases for packed columns also cover a greater polarity
range. The advantages of greater phase stability, giving less
bleed and improved EC sensitivity and MS background; and higher
resolution, giving a better separation of PEs from pesticides and
food matrix coextractants are two important reasons why bonded-
phase capillary columns are to be preferred over the packed column.
The reduced column bleed also enables temperature programming with
the EC detector. The multiple peaks of the various PCBs elute
throughout the range of the PEs on various gas chromatographic
columns. However, interferences from the PCBs are of little
concern as the Florisil chromatographic cleanup easily separates
the nonpolar PCBs from the relatively polar PEs. Some pesticides
will also be separated from the PEs by the Florisil cleanup. Again
the resolving power of the capillary column and the use of columns
of different polarity will enable identification and quantitation
of most PEs. Confirmation by MS is desirable.

EC detection is more widely used for PEs as well as pesticides
because it is more sensitive and selective than FID but GC-MS as
confirmation or as primary detection is becoming common. The
amount of PE injected to give the chromatograms in Figure 2 is
quite high, with the detector signal attenuated to give a noise-
free and relatively flat baseline. The electron capture response
of these phthalates at low attenuation is given in Table III. BBP
and DPP, with a phenyl group in the ester, give a detector response
about 10x that of the alkyl phthalates. Typically, the chlorinated
pesticides are detected at lower levels than BBP and DPP.

Table III. Electron Capture Response for Phthalate Esters

Phthalate	pg to give 20% F.S.D.
NDMP	380
DEP	640
DAP	100
DIBP	470
DNBP	410
BBP	40
DEHP	380
DPP	24
DNOP	320

Mass spectrometry is probably the most selective and desirable
detector for the GC determination of the PEs, although unequivocal
MS identification may not be possible. With electron impact
ionization, the PEs, except for DMP, give a strong characteristic
protonated phthalic anhydride ion base peak at m/z 149.

m/z 149

The higher esters, including DEHP, do not give a molecular ion (9). Methane chemical ionization, which was also largely unsuccessful in producing a protonated molecular ion for characterization of individual phthalates, also gives a base peak at m/z 149 for most PEs (10). Confirmative chromatography, using two different chromatographic phases is, therefore, highly desirable even with GC-MS detection.

Alternate Procedures

As described earlier, the ubiquitous presence of some PEs, especially DEHP, in the laboratory environment and in some analytical chemicals and materials requires extreme care to reduce the procedural blanks to a minimum. With low and consistent procedural blanks, reliable information concerning residue levels of the common PEs in foods can then be obtained. Preferred methods of analysis, therefore, must contain the least number of steps and use the minimum amounts of solvents, chemicals, adsorbents and other laboratory supplies in order to minimize such blank levels. The pesticide-type of procedure, as described earlier, does not meet these requirements. Therefore, it is desirable to consider alternate procedures which reduce or replace the steps of the pesticide-type procedure. The GC-EC or GC-MS determinative step is relatively straightforward and will remain the same for most procedures.

Steam Distillation. Steam distillation has been used to replace the extraction and/or partitioning steps outlined in Figure 1A and 1B. Such a procedure, using a modified apparatus originally designed to determine essential oils, has been described by Maruyama et al. (11) for the determination of PEs in foods. In their procedure 2 g butter, margarine or vegetable oil or 20 g of non-fatty food in a distilling flask are continuously refluxed in 100 mL of water. The aqueous distillate, containing trace PEs, condenses in the special distilling head, passes through 4 mL of heptane and returns to the flask. This 2 h distillation separates the PEs from the nonvolatile sample material. Most common organochlorine pesticides except for some of the pesticide metabolite p,p'-DDE are removed during reflux by the addition of 50 mg of zinc powder to the flask. The PCBs and some p,p'-DDE are recovered in the heptane. They are then separated from the PEs in the heptane by the addition of 0.5 g Florisil which absorbs the phthalates. The supernatant heptane, containing the PCBs and organochlorine pesticides including the p,p'-DDE metabolite, is removed and the Florisil rinsed with 4 mL hexane. 4 mL of an internal standard in acetone is added to recover the PEs from the Florisil for GC-EC determination. For the higher molecular weight PEs such as DEHP, it was necessary to add glycerol malate ester to facilitate the dispersion of oils and fats in the water and obtain good recoveries. The main advantages of the described steam distillation procedure is the potential to obtain low procedural blanks, as preextraction of fatty foods is avoided and only minimal amounts of organic solvents, water, Florisil and other chemicals are required. If desired, the distillation step can be used only to separate the PEs and any codistilled pesticides or PCBs from the sample. Subsequent separation of the steam-distilled components can then be carried out by other means. Distillation extracts, unlike those from simple extractions, will not contain non-volatile components, which could remain on the GC column. The main disadvantages are the 2 h distillation time and the requirement for a custom distillation apparatus, both of which limit sample throughput. Further studies to validate recovery and repeatability for the higher molecular weight PEs are required.

Sweep Codistillation (SCD). SCD using the Unitrex (Universal Trace Residue Extractor) apparatus has been used exclusively for the cleanup of fat samples for pesticide residue analysis (12,13). SCD can replace the partition step of Figure 1A and includes the Florisil fractionation, but with reduced Florisil and solvent requirements. A preliminary extraction to isolate the fat may be required. As lipid material is necessary for the proper operation of the SCD, blank lipid material such as vegetable oil could be added to extracts of non-fatty foods if cleanup by SCD is desired. The application of SCD to PE analysis has not been reported in the literature, however, preliminary work in our laboratory and by others suggests good recoveries of DEHP should be obtained.

The operation and components of the Unitrex SCD apparatus can be briefly described with reference to Figure 3. Up to 10 of these all-glass fractionation tubes with attached Florisil traps can be accommodated in the thermostatted heated block of the Unitrex apparatus. Nearly all of the tube is within the heated block as indicated by the heavy dotted lines in Figure 3. The nitrogen carrier gas is supplied to the fractionation tube by the septum head inlet and flows down the center tube, up between the 1.5 mm silanized glass beads which are packed in the annular space between the inner and outer tubes, and finally out through the Florisil sodium sulfate trap. 1.14 mL (1.0 g) of fat or oil, liquified by heating if necessary, is injected through the septum into the central tube. The lipid material passes down the inner tube and up the outer tube where the nitrogen flow spreads the lipid material over the surfaces of the glass beads and the inner and outer walls of the fractionation tube. The large surface area of the lipid allows a rapid transfer of the analyte between the gas and liquid phases. The volatilized pesticides, PCBs and PEs are then collected in the Florisil sodium sulfate trap. The SCD apparatus acts as a crude isothermal gas chromatograph with a mainly-triglyceride liquid phase, the analyte being initially dissolved throughout the liquid phase. Higher temperatures, increased flow rates or longer "elution" times are required for the less volatile analytes. Complete recovery of DEHP in 30 min may be achieved using a

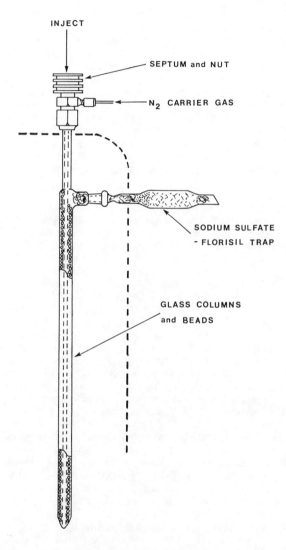

Figure 3. Sweep Codistillation Fractionation Tube and Trap.

nitrogen flow of 230 mL/min and a tube temperature of 250°C. After distillation the Florisil trap is removed and an eluting reservoir attached. The PCBs and organochlorine pesticides are eluted first, followed by the PEs. The 1 g of Florisil in the trap requires 10 mL of the appropriate eluant.

In summary, SCD rapidly and effectively separates the volatile phthalates from the non-volatile lipids. With the Unitrex system, 10 samples can be distilled simultaneously over 30 min. Typically, from 1 g of animal fat, only 1-2 mg of lipid material distills (12). SCD requires 1 g of Florisil and only 10 mL of eluant whereas the procedure outlined in Figure 1 requires 40 g and 300 mL, respectively, thus SCD has the potential for lowering procedural blank levels. The main disadvantage of SCD is that it is applicable only to 1 g fat samples. SCD is normally not applicable to non-fatty extracts unless fat is added. For fatty foods, the preextraction required to obtain the fat could increase the procedural blank level.

Gel Permeation Chromatography (GPC). The separation of pesticides and PEs from lipid material using GPC was first reported by Stalling (14). Burns et al. (15) later carried out a detailed study on a similar separation of fish lipids, four PEs, some common organochlorine pesticides and PCBs using a 32 x 2.5 cm column of BioBeads SX-3. Typically, ≤0.5 g of lipid is eluted with methylene chloride:cyclohexane (1:1); the lipids elute in the first 80 mL, the PEs in the next 20 mL and the chlorinated compounds in the last 40 mL. Although less than 10 mg of fish lipid coeluted with the PEs, subsequent sulfuric acid-impregnated alumina column cleanup was necessary to obtain interference-free extracts. The PEs were all eluted with 25 mL of 10% ethyl ether in hexane although a separate fraction containing only DEHP could be obtained. The GPC step gave 99% recovery of PEs. Overall, including a preliminary lipid extraction and the alumina column cleanup, recoveries from fish muscle samples spiked at 0.87 ppm ranged from 79.3 to 84.2%. From cod liver spiked at 7.8 ppm, the recovery as 82.4%. Quantitation was carried out by GC-EC using 2 packed columns.

GPC applied to lipid materials can advantageously replace the liquid-liquid partition step and the Florisil cleanup of Figure 1. Automated systems, which can sequentially process more than 20 0.5 g lipid extract samples, require about 30 min to fractionate for each sample. GPC equipment, however, is not common equipment in most laboratories.

High Pressure Liquid Chromatography (HPLC). HPLC has occasionally been used to determine PEs in environmental samples such as water or sediment. The PEs are extracted into hexane, concentrated, and determined by normal phase HPLC. HPLC has limited application to determine PEs in fatty foods. One report (16) describes the injection of a diluted oil onto a reverse phase column with UV detection. 50 ppm of DEHP could readily be detected. Compared to the low pg amounts detected by GC-EC or GC-MS, the UV detection in HPLC is limited to low ng quantities, and, as a result, HPLC methods are inherently less sensitive. Trace enrichment techniques applied to some aqueous samples may be an exception. Perhaps the

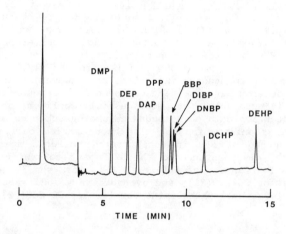

Figure 4. HPLC Chromatograms of Phthalate Esters on Supelcosil RP-18 by Gradient Elution. See text for details.

most useful application of HPLC is the screening of packaging materials for PEs. Extracts of the packaging material can be readily analyzed by reverse phase gradient HPLC. A chromatogram of 9 PEs separated in the author's laboratory on a 150 x 4.6 mm i.d. Supelcosil LC-18 column using a non-linear 0 to 100% acetonitrile in water gradient is shown in Figure 4. The PEs, about 200 ng each, were detected at 274 nm (0.04 AUFS). The order of elution is similar to GC. As the molecular weight increases the volatility and the polarity decrease giving longer retention times for GC and HPLC, respectively.

Supplementary Cleanup and Confirmation. Cleanup of the extract is indicated if the PE peaks detected by GC-EC are obscured by co-eluting sample components. Column chromatography on alumina (15) or Florisil (17) may be used to reduce such interferences. Partition of the PEs from the hexane extract solution into sulfuric acid has been used by several workers to isolate PEs (17,18). Chemical means, such as permanganate oxidation, zinc reduction, and sulfuric acid decomposition has also been employed for cleanup (19).

GC-MS is probably the most useful confirmation technique although retention data on a second chromatographic column has been widely used. Confirmation has also been carried out by examining the fluorescence of PEs in concentrated sulfuric acid (20), by observing the disappearance of the DEHP peak following hydrolysis (14), by derivatization of DBP and DEP to form the 3- and 4-nitro-derivatives for GC-EC (21), and by hydrolysis and reaction to form the N-(2-chloroethyl) phthalimide for further GC-EC (22).

Summary

Phthalate esters can readily be determined in fatty or non-fatty foods by well-established pesticide procedures. Stringent protocol, however, is required to reduce procedural blank levels and associated detection limits of the common PEs. These background PEs arise from reagents, equipment, and the laboratory environment. Alternate procedures, which minimize the contribution from these sources, including sweep codistillation, GPC, HPLC and possibly steam distillation, should be studied in more detail. Supplementary cleanup and confirmation steps improve detection and reliability. GC-EC or GC-MS is normally used for quantitation.

Literature Cited

1.   Anon. Mod. Plast. 1986, 16, 81.
2.   Carcinogenesis Bioassay of Di(2-ethylexyl) Phthalate (CAS No. 171-81-7) in F334 Rats and B6C3FI Mice (Feed Study), National Toxicology Program, Technical Report Series No. 217; NIH Publ. No. 82-1773; Research Triangle Park, NC.
3.   Giam, C.S.; Chan, H.S.; Neff, G.S. Anal. Chem. 1975, 47, 2225-2229.
4.   Ishida, M.; Suyama, K.; Adachi, S. J. Chromatogr. 1980, 189, 421-424.
5.   Krauskopf, L.G. In Encyclopedia of PVC; Nass, L.I., Ed.; Marcel Dekker: New York, 1976; Vol. 1, p 520.

6. Analytical Methods for Pesticide Residues in Foods, Health and Welfare Canada, Canadian Government Publishing Centre: Ottawa, Canada, 1986.

7. Pesticide Analytical Manual, U.S. Department of Health and Human Services, Food and Drug Administration, 1986.

8. Group, E.F. Jr. Environ. Health Perspect. 1986, 65, 337-340.

9. Stalling, D.L.; Hogan, J.W.; Johnson, J.L. Environ. Health Perspect. 1979, 3, 159-173.

10. Cairns, T.; Chiu, K.S.; Siegmund, E.G.; Williamson, B.; Fisher, J.G. Biomed. Environ. Mass Spectrom. 1986, 13, 357-360.

11. Maruyama, T.; Ushigusa, T.; Kanematsu, H.; Niiya, I.; Imamura, M. Shokuhin Eiseigaku Zasshi 1977, 18, 244-251.

12. Luke, B.G.; Richards, J.C.; Dawes, E.F. J. Assoc. Off. Anal. Chem. 1984, 67, 295-298.

13. Brown, R.L.; Farmer, C.N.; Millar, R.G. J. Assoc. Off. Anal. Chem. 1987, 70, 442-445.

14. Stalling, D.L.; Tindle, R.C.; Johnson, J.L. J. Assoc. Off. Anal. Chem. 1972, 55, 32-38.

15. Burns, B.G.; Musial, C.J.; Uthe, J.F. J. Assoc. Off. Anal. Chem. 1981, 64, 282-286.

16. Niebergall, H.; Hartmann, M. Lebensm. Wiss. Technol. 1983, 16, 254-262.

17. Saito, Y.; Takeda, M.; Uchiyama, M. Shokuhin Eiseigaku Zasshi 1976, 17, 170-175.

18. Thuren, A.; Sodergren, A. Intern. J. Environ. Anal. Chem. 1987, 28, 309-315.

19. Suzuki, T.; Ishikawa, K.; Sato, N.; Sakai, K.-I.; J. Assoc. Off. Anal. Chem. 1979, 62, 689-694.

20. Zitko, V. Intern. J. Environ. Anal. Chem. 1973, 2, 241-252.

21. Ishikawa, K.; Sato, N.; Takatsuki, K.; Sakai, K. Eisei Kagaku 1977, 23, 175-179.

22. Giam, C.S.; Chan, H.S.; Hammargren, T.F.; Neff, G.S. Anal. Chem. 1976, 48, 78-80.

RECEIVED November 5, 1987

Chapter 11

Recent Advances in Analytical Methods for Determination of Migrants

Henry C. Hollifield, Roger C. Snyder, Timothy P. McNeal, and Thomas Fazio

Center for Food Safety and Applied Nutrition, Food and Drug Administration, Washington, DC 20204

Within the past several years, new developments in
analytical technology have enabled a rapid advance in
the laboratory investigation of polymer monomers,
oligomers, and adjuvants, as well as the determination
of their migration into foods and food-simulating
solvents. In some cases, it is possible to measure
low parts per billion levels in the polymer and parts
per trillion levels in food simulants. This paper
describes some of the newer approaches and
applications developed to survey consumer products and
to evaluate proposed new materials. Included are
discussions of (a) a new cell for assessing migration
into simulating liquids, (b) mathematical modeling of
migration data and application to selection of a fatty
food simulant, (c) a multiresidue approach for
determining packaging-derived volatile residues in
aqueous foods, and (d) use of combined size-exclusion
and high-performance liquid chromatographic residue
methods.

The diffusion of chemicals through polymers and their migration into
foods are phenomena that are of interest to the Food and Drug
Administration (FDA) in its regulation of polymeric food-packaging
materials. In 1958 Congress amended the Food, Drug, and Cosmetics
Act, defining indirect additives to foods and requiring their
regulation. Since the passage of the 1958 Food Additives Amendment,
FDA has required migration studies to provide exposure estimates for
newly regulated indirect additives (1).
Within the past several years, new developments in analytical
technology have enabled important advances in the laboratory
investigation of polymeric monomers, oligomers, and adjuvants and
the determination of their migration into foods and food-simulating
liquids (FSL). In some cases, it is possible to measure parts per
billion levels in the polymer and to determine parts per trillion
levels in the food-simulating solvents. Some of the more recent

developments include (a) a new migration cell for two-sided extraction of polymeric resin compounds, (b) new fatty food simulants, (c) combined analytical techniques for trace level measurements, and (d) multiresidue procedures. This paper discusses some of these advances in which FDA has actively participated.

In 1976, FDA awarded a multiyear contract to Arthur D. Little to assess the suitability of traditional food simulants such as water, 3% acetic acid, heptane, and corn oil (2). That same year the National Bureau of Standards (NBS) entered into an interagency agreement with FDA to investigate the factors controlling migration (3). Both of these efforts used radiolabeled additives and/or polymers. From these studies came a number of important results. For example, a new, more efficient two-sided extraction cell was developed, mathematical models were used to describe the migration results, and alternative food-simulating solvents were investigated.

Evaluation of the New Migration Cell

FDA subsequently followed up these studies with its own evaluation of the new cell (4). A new BASIC computer program was used to fit the observed migration data to the proposed mathematical model. The cell was applied to the study of improved fatty food simulants by using a styrene/polystyrene system. This choice was based on the fact that typical residual styrene levels (500-3000 ppm) in crystalline polystyrene permit a "cold" (unlabeled) study. Also, the volatility of styrene is high enough to be used to evaluate the integrity of the cell.

The migration cell is illustrated in Figure 1. It consists of a 23-mL glass screw-cap vial and polymer discs on a rack assembly. The vials are filled with the desired test solvent and sealed with the Miniert Teflon cap containing a slide valve that permits repeated sampling and resealing. Each vial contains a constant number of discs stacked on a wire and separated by 3-mm glass beads. For 14 polystyrene discs in 22 mL of food-simulating solvent, the volume-to-surface ratio can be accurately determined (0.73 mL/sq. cm in our studies). Assembled cells are held in darkness at the desired test temperature in a water bath. The water bath provides mild agitation of the cell at the rate of 60 cpm with a stroke of 4 cm.

Polystyrene discs for migration studies are prepared by pressing polystyrene pellets in a template at 400°F (204°C) under a 15,000 psi force for 30 s. The resultant circular discs are of constant diameter, thickness, and weight. They are characterized by size-exclusion chromatography (SEC), and the residual styrene monomer level is determined by high-performance liquid chromatography (HPLC). A small hole is drilled in the center of each disc so that the discs may be stacked on the wire support.

Reproducibility of data generated by this technique was evaluated by preparing six identical migration cells, with water at 40°C as the contact solvent. The solvent in each cell was periodically analyzed for styrene concentration over a period of 2 months. The overall relative standard deviation for these determinations was ±2%, indicating the excellent repeatability obtainable by this migration cell technique.

Styrene migration into the FSL was monitored by reversed-phase
HPLC. A 5-μm C_8 Zorbax column interfaced with a fixed wavelength
(254 nm) UV detector was employed. A mobile phase of 75%
acetonitrile in water at a flow rate of 1 mL/min proved adequate to
separate styrene from all interferences. Analysis of all the
various solvent systems on the same column was possible by use of a
Brownlee loop column, which consisted of a 3.7-cm long cartridge
packed with 10-μm C_8 reversed-phase packing material. The loop
column is mounted in the injector valve where a fixed volume
injector loop would normally be located. Since the loop column
retains oily sample components and allows materials of lower
molecular weight to pass through first, it permits direct injection
of oily as well as aqueous FSL. Oily materials are removed from the
loop column between sample injections by flushing the loop column
with tetrahydrofuran. Styrene migration from polystyrene discs into
the following solvents was monitored at 40°C: water; 8, 20, 50, and
100% ethanol; HB-307 (synthetic triglyceride); corn oil; heptane;
hexadecane; decanol; and 3% acetic acid. The migration cells were
periodically sampled (25 μL), and the concentration of styrene
was determined by external calibration. The limit of
quantitation of the procedure is 10 μg styrene/L. Styrene
migration, expressed as μg/sq. cm vs. square root of time
(s), into a variety of solvents at 40°C is shown in Figure 2.
This figure shows that at 40°C, styrene migrated at similar rates
into 20% ethanol, corn oil, and HB-307. Figure 2 reveals that
varying the ethanol concentration from 0 to 50% in water caused the
rate of styrene migration to vary over one order of magnitude.
Ethanol-water mixtures, therefore, would seem well suited as low
temperature fatty food simulants for a wide variety of foods.

Mathematical Modeling

Data obtained from migration studies can be interpreted in
terms of simplified mathematical models developed jointly by
NBS and Arthur D. Little. For example, by fitting the data
generated in the above study to the migration models, several
observations can be made about this styrene-polystyrene
system. For short periods of time (<8 h), the migration of
styrene from the polystyrene discs was observed to be
Fickian, that is, styrene migrated at a rate linearly
dependent on the square root of time. This is expected
because (a) only a small percentage of the total available
styrene migrated and, therefore, the styrene concentration in
the polymer was essentially constant, and (b) over short
periods of time the solvent acted as an infinite sink and,
therefore, the concentration in the solvent could be treated
as equal to zero. Under these boundary conditions the
migration could be expressed by Equation 1:

$$M_t = 2C_p(D_p t/\pi)^{1/2} \tag{1}$$

where M_t = migration, μg/sq. cm; t = time, s; D_p = diffusion
coefficient in the polymer, sq. cm/s; and C_p = initial concentration
in the polymer, μg/cu. cm. However, for longer periods of time, an

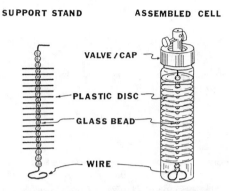

Figure 1. Migration cell consisting of polystyrene discs, glass beads, copper wire, 23-mL glass vial, and valve cap. (Reproduced with permission from Ref. 4. Copyright 1985 J. Assoc. Off. Anal. Chem.)

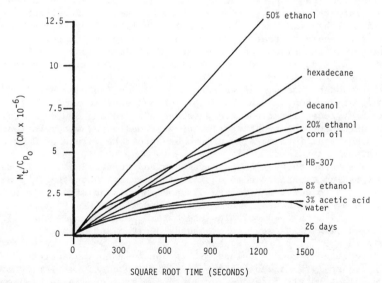

Figure 2. Styrene migration (μg/sq. cm)/(μg/cu. cm) into various solvents at 40°C as a function of time ($s^{1/2}$). (Reproduced with permission from Ref. 4. Copyright 1985 J. Assoc. Off. Anal. Chem.)

equilibrium partitioning (K) exists between the concentration of the styrene in the polystyrene discs (C_p) and the concentration in the solvent (C_s). The solvent must, therefore, be treated as having a finite volume per unit area of polymer surface (a). Also, the geometry of the system (u) must be taken into consideration. Under these boundary conditions the migration is given by Equation 2 (5,6):

$$M_t = C_p aK(1 - e^{Z^2} \text{ erfc } Z) \qquad (2)$$

where

$$Z = (D_p t)^{1/2}/aK$$

$$\text{erfc} = 2/(\pi)^{1/2} \int_z^\infty e^{-u^2} du$$

Because these are simplified models, they do not take a number of factors into account. For example, blooming of adjuvants on the surface of the polymer, temperature variations, solvent viscosity, and solvent penetration of the polymer can all result in deviations from these model cases. In some cases the models can be modified to account for some factors such as solvent viscosity. An additional diffusion coefficient is required if the food simulant is a solid or very viscous. In our work all were nonviscous liquids and migration could be modeled by either Equation 1 or 2. Calculated D_p and K values for styrene migration at 40°C for a variety of solvents are listed in Table I. Use of Equation 1 to calculate D_p is the simplest, requiring that migration be monitored only a few hours. However, surface bloom can cause very large deviations. Using Equation 2 to determine D_p requires monitoring the migration over much longer time intervals. Also, the K value between polymer and solvent must be known. If an accurate K value is to be obtained, the migration should be allowed to approach equilibrium. In the case of styrene migration into water at 40°C, about 300 days would be required for the concentration of styrene in the water to reach 90% of the equilibrium value on the basis of the values of D_p and K ultimately determined.

To shorten these time requirements and to avoid having to take the migration experiments to equilibrium, an iterative BASIC program was written to solve for D_p and K, based on six to eight data points obtained over 2-3 weeks. The program, described in detail in reference (4), is used to calculate both D_p and K values that fit the linear line-fit curve generated by a given set of migration data to within a 98% fit. On the basis of the slope of this curve and the calculated D_p and K values, it is possible to extrapolate predicted analyte concentrations if migration continues for extended times. Thus, shelf life and equilibrium migration can be predicted.

Table I shows an average D_p of 3×10^{-13} sq. cm/s for styrene from crystalline polystyrene, except for 50 and 100% ethanol. This value is in good agreement with D_p values reported by others. The D_p values calculated for 50 and 100% ethanol are not considered the true "inherent D_p" values but rather the "effective" diffusion coefficients and reflect a greatly increased diffusion rate due to

Table I. Styrene Diffusion Coefficients and Partition
Coefficient from Polystyrene at 40°C[a]

Solvent	D[b] (x 10^{-13} sq. cm/s)	D[c] (x 10^{-13} sq. cm/s)	K (x 10^{-3})
Water	1	1-3	0.3[d]
3% Acetic acid	1	2	0.3[d]
8% Ethanol	2	2	0.5[d]
20% Ethanol	4	7	1.4[d]
50% Ethanol	12	30	30
100% Ethanol	50	150	170
Corn oil	3	2	5
HB-307	4	7	1
Hexadecane	3	3	65
Decanol	3	3	4

[a] Reprinted with permission from Ref. 4.
[b] From Equation 1; 24-h data point.
[c] From Equation 2.
[d] Confirmed by reverse migration experiments.

solvent penetration. Reverse migration experiments were found to be
a fairly rapid means of checking the K values calculated by using
the iterative BASIC program.

There are several advantages to being able to predict migration
performance of a given indirect additive in the presence of various
FSL. Obviously, resources can be saved by using short-term
migration studies instead of carrying out every study to
equilibrium. A less obvious advantage is that the mathematical
models and the supporting migration studies conducted by NBS, Arthur
D. Little, and the Indirect Additives Laboratory have given FDA a
solid foundation for evaluating the consistency and probable
reliability of submitted petition data. There is greater confidence
in the reliability of (a) short-term migration studies, 10 days at
120°F vs. 30 days, and (b) projections based on short-term data.
These kinds of studies have given the agency a sound scientific
basis for recommending the most expeditious migration studies to
meet the regulatory requirements for indirect food additive petition
approval.

Multiresidue Headspace Procedures Employing Computerized Data Reduction

Just as computers have become a valuable asset in mathematical
modeling of migration studies, they have also become indispensable
in multiresidue methods for the measurement of food packaging-
derived volatiles in food and containers. For example, we have used
the computer and the method of standard additions to quantify
multiple residues in various aqueous foods including milk and infant
formula (7, McNeal, T. P.; Hollifield, H. C., in preparation). The
procedure combines the sensitivity and repeatability of automated

headspace sampling, the resolving power of capillary gas
chromatography, and the speed of automated data reduction to handle
the many replicate measurements required. A typical analysis of a
given sample may require up to 10 chromatograms, including controls
and duplicates.

This procedure is most useful in screening large numbers of
laboratory samples for several components at once. For example, to
screen milk for potential contamination during collection,
processing, or storage, aliquots are screened for possible
contaminants. Based on the relative amounts of each contaminant
found in the screening procedure, standards are prepared for the
components of interest and are used to fortify test samples. Flame
ionization detection (FID) chromatograms of milk and fortified milk
are shown in Figure 3, illustrating the screening for acetone,
methyl ethyl ketone (MEK), benzene, and toluene. Quantitative
results are obtained from a linear regression plot of peak data
obtained following standard additions. Acetone, MEK, hexane,
benzene, trichloroethylene, perchloroethylene, cyclohexane, and
heptane were some of the analytes used to test the procedure. Most
of the analytes tested have retention indexes between 500 and 800.

For this analysis we have typically employed 10-mL sample
aliquots in septum-sealed headspace vials heated at 90°C for 30 min
in a Perkin-Elmer Model F-42 or Model HS100 headspace analyzer and
Perkin-Elmer Sigma 2000 gas chromatograph equipped with a capillary
injection system and a DB-1 nonpolar column. Raw and reduced data
are collected by an IBM Model 9000 computer with the IBM
Chromatography Applications Package software. A custom BASIC
computer linear-regression program plots the chromatographic peak
responses of selected components vs. fortification levels for the
determination of their concentrations.

In addition to the concentration of each component, a
statistical evaluation of the data is also given. The correlation
coefficient, standard deviation, 95% confidence interval, and lower
limit of reliable measurement are calculated for each set of data.
These factors provide an indication of the precision and reliability
of the determination. Raw data are archived on floppy computer
discs for future use.

The method has been successfully tested with selected chemicals
on both normal bore (0.2 mm id) and wide bore (0.5 mm id) capillary
columns. A comparison of the relative responses of several of these
analytes on both columns is given in Table II. Note that the ratio
of the relative responses on the wide bore column to those on the
normal bore column increases from 19:1 for acrylonitrile to 32:1 for
tetrahydrofuran and benzene. For greater sensitivity, the large
bore column is the better choice although some loss of resolution
occurs. The last column of the table gives the lower limits of
reliable quantitation on large bore capillary columns for our
headspace system using FID. It is expected that sensitivities for
some of the analytes can be improved further by altering headspace
parameters and using selective detectors.

In addition to the static headspace technique described above,
we also use other dynamic headspace sampling methods. For some
time, we have relied on a research purge and trap headspace approach

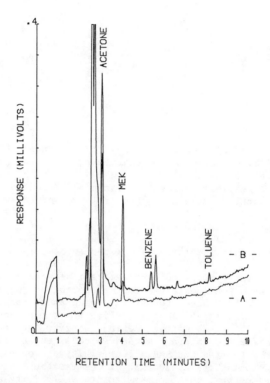

Figure 3. FID chromatograms obtained by screening milk samples for selected chemicals. Fortification levels in ng/mL were acetone (450), methyl ethyl ketone (450), benzene (16), and toluene (16). A = unfortified milk; B = fortified milk.

Table II. Comparison of Normal Bore and Megabore FSOT
Normalized Relative Response Factor Ratios, Using
the F-42 Automated Headspace Analyzer

Compound	Relative Response Ratio[a]	LRM[b] Milk (Megabore) (ppb)
Acrylonitrile	19:1	2
Methylene chloride	28:1	–
Vinylidene chloride	18:1	4
Methyl ethyl ketone	26:1	3
Chloroform	29:1	7
Tetrahydrofuran	32:1	2
Benzene	32:1	1
Toluene	27:1	–

[a]Response factors (RF) represent microvolts response/ppb analyte added to 2 mL deionized water. Relative response factor ratio = (RF megabore/RF normal bore).
[b]LRM = Limit of reliable measurement; at the 95% confidence level, it is approximately 1.96 sigma (standard error of the regression line).

to monitor vinyl chloride in polyvinyl choloride polymers at low parts per billion levels (8). Although this approach has parts per trillion sensitivity, it is cumbersome, tedious, and slow. It is not desirable as a routine method for monitoring or quality control.

There is considerable interest in having an automated, easy to use, quality control method that is reliable for the vinyl chloride monomer. Method D4443-84 of the American Society for Testing and Materials is a static headspace method which is relatively easy to use; it can be automated and has a sensitivity approaching this level (9). A statistical evaluation of the existing data supporting this method raises some questions as to its reliability at the low parts per billion range. The method sensitivity appears to vary with the polymer matrix analyzed. This method is presently under reevaluation using various polyvinyl chloride (PVC) polymers to unambiguously determine its sensitivity and reliability.

In spite of this apparent move away from dynamic headspace procedures for polymeric residues and migrants, purge and trap methods are the most promising approach when very low quantitative limits are desired. One such example is the determination of benzene in polypropylene. Headspace analysis of the trapping solutions on large bore capillary columns using FID has a demonstrated sensitivity down to 20 ppb benzene.

Another recent advance in the determination of migrants used in our laboratory has been the use of SEC for cleanup and isolation of polymeric residues. We first used SEC to separate phthalates and adipates from solvent extracts of PVC films collected for market surveys. In this analysis, the analytes are extracted from PVC films with toluene, and the extract is subsequently chromatographed on a suitable SEC column using a refractive index detector. Fractions containing plasticizers are collected and determined by gas chromatography using external standards for quantitation.

Similar approaches have been used with HPLC procedures for the separation and quantitation of styrene in polystyrene and for dimethyl terephthalate and terephthalic acid residues in polyethylene terephthalate polymers. Concentration steps may be included following SEC to achieve increased sensitivity.

In conclusion, the explosive growth of plastics technology provides fertile soil for many possible chemical interactions between foods and new polymeric food packaging materials. The rapid growth of new retortable, ovenable, and microwavable packages reflects the need for studies in high temperature applications in addition to continued efforts to detail lower temperature food-package interactions. The analytical challenges this offers are formidable and exciting. It will call for our continued best efforts to develop new and innovative advances in analytical methods for the determination of migrants.

Literature Cited

1. FDA Guidelines for Chemistry and Technology Requirements of Indirect Food Additive Petitions, Food and Drug Administration: Washington, DC, March 1976.
2. Arthur D. Little, A Study of Indirect Food Additive Migration, Final Summary Report, July 1983, FDA Contract No. 223-77-2630.
3. Migration of Low Molecular Weight Additives in Polyolefins and Copolymers, U.S. Department of Commerce, National Bureau of Standards: Washington, DC, March 1982, NBS IR-2472.
4. Snyder, R. C.; Breder, C. V. J. Assoc. Off. Anal. Chem. 1985, 68, 770-5.
5. Carslaw, H. S.; Jaeger, J. C. Conductance of Heat in Solids, 2nd ed.; Clarendon Press: Oxford, UK, 1959.
6. Reid, R. C.; Sidman, K. R.; Schwope, A. D.; Till, D. E. Ind. Eng. Chem. Prod. Res. Dev. 1980, 19, 580-7.
7. McNeal, T. P., Hollifield, H. C. Pittsburgh Conference & Exposition on Analytical Chemistry and Applied Spectroscopy, Atlantic City, NJ; Abstract 651.
8. Dennison, J. L.; Breder, C. V.; McNeal, T.; Snyder, R. C.; Roach, J. A.; Sphon, J. A. J. Assoc. Off. Anal. Chem. 1978, 61, 813-19.
9. 1986 Annual Book of ASTM Standards, D-4443-84, American Society for Testing and Materials: Philadelphia, PA, 1986.

RECEIVED September 24, 1987

Chapter 12

Migration and Formation of *N*–Nitrosamines from Food Contact Materials

Nrisinha P. Sen

Food Research Division, Bureau of Chemical Safety, Food Directorate, Health Protection Branch, Ottawa, Ontario K1A 0L2, Canada

Recent research has focused our attention on the occurrence of N-nitrosamines, many of which are potent carcinogens, and secondary amines, in food contact materials. In addition to the possibility of migration of these chemicals to foods, there is also the possibility that the migrated amines, after ingestion, may form additional amounts of nitrosamines, either in vivo in the human stomach, or during processing or cooking of foods containing nitrite as an additive (e.g., cured meats). This paper reviews the topic with particular emphasis on paper-, wax-, and rubber-based food contact materials. The available evidence indicates that certain nitrosamines may migrate to or form in foods which come in contact with rubber based products such as baby bottle rubber nipples and elastic rubber nettings.

N-Nitrosamines are a group of environmental carcinogens that have been detected in a wide variety of consumer products such as pesticide formulations; cigarette and other tobacco products; cosmetics; rubber products; and in various foods and beverages, especially fried bacon, cured meats and beer. In foods, these compounds are formed mainly as a result of the interaction of naturally occurring amines in the foods, and nitrite additives (e.g., in bacon) or nitrogen oxide gases (e.g., in hot flue gases used for drying malt) used in the processing of the products. Since most of the nitrosamines detected in foods are potent carcinogens, these findings have aroused a great deal of concern among health officials and consumers throughout the world. Details of these findings have been published in several recent reviews (1-3).

Recent research has shown that some of the nitrosamines detected in foods may originate or form from various food contact materials such as packaging papers, waxed containers, and elastic rubber nettings. In these cases, the food contact materials, not the foods,

0097–6156/88/0365–0146$06.00/0
Published 1988 American Chemical Society

are the source of the nitrosatable amines or amine derivatives. The concern, here, appears to be three-fold. First, the nitrosamines from these products can migrate to foods that come in contact with them for a prolonged period. Secondly, the amines, which are usually present in much higher concentrations than the corresponding nitrosamines, can also migrate to foods and form additional amounts of nitrosamines during processing or cooking. And thirdly, the migrated amines, after ingestion, can also form nitrosamines in vivo in the acidic environment of the human stomach due to interaction with salivary or ingested (e.g., through cured meats) nitrite.

Table I summarizes the type of food contact materials that have been implicated in the contamination of foods with nitrosamines and nitrosatable amines. The available evidence suggests that N–nitrosomorpholine (NMOR) and morpholine (MOR) are present as contaminants in paper-based packaging materials and liquid waxes, whereas rubber-

Table I. Food Contact Materials Reported to Contain N–Nitrosamines

Type	Use pattern	Nitrosamines detected
Paper-based products including waxed papers	Used for packaging dry as well as moist foods	NMOR
Liquid waxes	As a coating on fruits and vegetables	NMOR
Rubber-based products	Baby bottle rubber nipples	N–Nitrosodimethylamine (NDMA) N–Nitrosodiethylamine (NDEA) N–Nitrosodi-n-butylamine (NDBA) N–Nitrosopiperidine (NPIP) NMOR N–Nitrosomethylphenylamine (NMPhA) N–Nitrosoethylphenylamine (NEPhA) N–Nitrosodiphenylamine (NDPhA)
	Elastic rubber nettings used for packaging cured meats	NDEA NPIP NDBA N–Nitrosodibenzylamine (NDBzA)

based products contain a variety of nitrosamines and the corresponding amine derivatives. Although only limited information is available, some interesting observations have been made. An attempt will
be made, in this review, to summarize these findings and discuss
their significance with regard to the possible health hazard to man.

N-Nitrosamines and Precursor Amines in Paper-Based Food Contact Materials

The first evidence of the presence of NMOR in paper-based packaging
materials was reported independently by Hoffmann et al. (4) and
Hotchkiss and Vecchio (5). While investigating the origin of NMOR in
snuff, Hoffmann et al. (4) observed that the wax coating used on the
snuff containers was the source of the NMOR contamination. Since MOR
is widely used as a solvent in most wax formulations and since most
commercial MOR is contaminated with NMOR (6), the finding of NMOR in
the wax-coated snuff containers was not entirely unexpected. The
above researchers (4) also analyzed a variety of waxed containers
commonly used for packaging dairy products for the presence of both
NMOR and MOR (Table II). Fairly high levels of MOR and traces of
NMOR were detected in such containers. The respective dairy
products, packaged in these containers, also contained significant
levels of MOR and traces of NMOR (Table II).

On careful examination of the data (Table II) it appears that
the concentration of NMOR in six of the dairy foods was higher than
that present in the corresponding containers. If NMOR in these foods
had solely originated from the containers the reverse should have
been true as in the case for MOR, which was present in much higher
concentrations in the containers than in the respective foods. The

Table II. NMOR and MOR in Food and Food Containers (ppb)

Sample	Food		Container	
	NMOR	MOR	NMOR	MOR
Butter	3.2	58	1.9	220
Cream cheese	0.9	77	N[a]	680
Yogurt	N	38	N	3,060
Cottage cheese	0.4	44	5.4	17,200
Frozen peas and carrots	N	26	3.1	57
Cheese (semi-soft)[b]	3.3	8.7	N	26
Cheese (semi-soft)[b]	3.1	9.7	1.6	25
Cheese (semi-soft)[b]	0.7	4.9	1.2	22
Cheese (semi-soft)[b]	1.4	8.0	N	132
Gouda	1.6	35	N	35

[a] N = None detected (<0.2 ppb)
[b] Each was from a different country
(Source: Reproduced with permission from ref. 4. Copyright 1982 Cold Spring
Harbor Laboratory.)

reason for finding higher levels of NMOR in some of the foods is not clear.

Since MOR is widely used as a corrosion inhibitor in boiler feed water and since large amounts of steam and water are used in the manufacture of paper and paperboard packagings, there is a possibility of finding both MOR and NMOR as contaminants in such products (5). This led Hotchkiss and Vecchio (5) to analyze a variety of food grade paper and paperboard packaging materials for the presence of the above-mentioned contaminants. Most of these paper products were of the type that are used for packaging dry foods such as oats, cereal, rice, pasta, sugar, flour, salt, and cornmeal. Of the 34 samples analyzed, nine contained traces (1 to 33 ppb) of NMOR (unconfirmed by mass spectrometry); the rest were negative. The above researchers also analyzed six each of the NMOR-positive and NMOR-negative samples for their MOR contents. All contained fairly high levels of MOR (Table III). In some cases, the levels of NMOR correlated with those of MOR.

Table III. MOR and NMOR Contents of Food Grade Paper and Paperboard Packaging Materials

Sample	MOR (ppb)	NMOR (ppb)
A	426	15.1
B	223	trace[a]
C	560	3.6
D	347	8.9
E	812	13.1
F	238	trace
$\bar{X}$	434	
G	98	N[b]
H	132	N
I	329	N
J	445	N
K	113	N
L	842	N
$\bar{X}$	327	

[a] trace (<3 ppb)
[b] N = None detected
(Source: Reproduced with permission from ref. 5. Copyright 1983 Institute of Food Technologists.)

Hotchkiss and Vecchio (5) also investigated the possible migration of these chemicals from packaging papers to foods. Traces (~1
ppb) of NMOR were observed in flour samples taken closest to the bag
wall which itself contained about 33 ppb NMOR. Homogenized flour
from another bag was found to contain 18 ppb MOR but no NMOR,
suggesting migration of MOR from the bag to the flour. When pieces
of paper and paperboard packagings containing NMOR were incubated for
72 hr at 100°C with flour or pasta in a closed container, traces of
NMOR migrated from the packagings to these foods. These workers
emphasized, however, that since the conditions used for the above
migration experiments were quite drastic, the data should be interpreted with caution. Further work is obviously needed to determine
migration patterns for these chemicals from packaging materials to
foods under normal storage conditions. It would be advisable to
replace MOR as a corrosion inhibitor in boiler water with an amine
that does not form a stable N-nitroso derivative (5).

More recently, Sen and Baddoo (7) reported further evidence of
migration of NMOR from packaging papers to margarine. Margarines and
a variety of wrappings (parchment paper, specially coated paper,
waxed paper, aluminum-backed paper) were analyzed for their NMOR
content. NMOR ranging from 5 to 73 ng per wrapping was detected in
some of the wrappings. Furthermore, margarines taken from the outer
~5 mm layer of sample packaged in NMOR-positive wrappings were mostly
positive for NMOR, while those taken from the center of the margarine
blocks (1 lb) were always negative. This suggested migration of NMOR
from the wrappings to the outer layer of the margarine blocks. Not
all margarines wrapped in NMOR-positive wrappings were, however,
contaminated with NMOR. In fact, the brand of paper containing the
highest levels of NMOR (62 to 73 ng/wrapper) did not transfer any
NMOR to margarine. This was attributed to the presence of a special
plastic coating on the innerside of these wrappings that prevented
NMOR migration. No NMOR was detected in margarines packaged in
aluminum-backed paper or plastic tubs.

MOR and NMOR Contamination of Fruits from Liquid Waxes

The application of liquid wax as an edible protective coating on
fruits and vegetables is an old practice (8). Its main purpose is to
prevent moisture loss and thus extend shelf life of fruits and
vegetables (8). Since MOR and fatty acid derivatives of MOR (e.g.,
morpholine oleate) are widely used as solvents or emulsifying agents
for waxes (9), such a practice can inadvertently contaminate wax-
coated fruits and vegetables with MOR and NMOR. Unfortunately, not a
great deal of information is available in this regard. Only very
recently, Sen and Baddoo (10) investigated the possible presence of
MOR and NMOR in apples coated with liquid waxes. As expected, up to
3.8% MOR and 140 to 670 ppb of NMOR were detected in 9 samples of
liquid waxes. In apples coated with such waxes, 0.25 to 7.7 ppm of
MOR but no NMOR were detected. Neither MOR nor NMOR were detected in
uncoated apples.

Since MOR is an easily nitrosatable amine (11), the above researchers (10) also investigated the possibility of the formation of NMOR in the human stomach following ingestion of wax-coated apples. In vitro incubation studies with wax-coated apple homogenates (incubated with 10 ppm nitrite solution at pH 3.4 for 2 hrs at 37°C) gave mostly negative results, whereas equivalent amounts of MOR under identical conditions in the absence of apple, produced traces of NMOR. The negative results with apples were attributed to the presence of naturally occurring N-nitrosation inhibitors (e.g., phenolic compounds, ascorbic acid) in these fruits (11-13). In view of these findings, the use of liquid waxes on fruit appears to be of little health hazard significance to man, at least in the context of NMOR contamination or its formation.

Nitrosamines in Rubber-Based Food Contact Materials

Recent research has provided conclusive evidence for the presence of various volatile nitrosamines in a variety of rubber products such as tires and rubber tubes, rubber gloves, baby bottle rubber nipples and infant pacifiers, and in some medical devices (14-19). The nitrosamines detected in these products are listed in Table I. The best evidence indicates that these nitrosamines are formed as a result of the interaction of various vulcanization accelerators (amine derivatives) in the rubber and nitrogen oxides from air. Some organic nitro and nitroso compounds (e.g., N-nitrosodiphenylamine) used in the manufacture of rubber can also act as nitrosating agents. The reader is advised to consult a recent review (16) on the subject for further details.

The finding of nitrosamines in baby bottle rubber nipples raised a great deal of concern because of the possible migration of these compounds to infant formulas and into babies' saliva during their normal usage. Furthermore, many mothers sterilize infant formulas in bottles with rubber nipples placed in them, and often store them together for a prolonged period in the refrigerator. Therefore, the opportunity for migration of nitrosamines as well as of the corresponding amine derivatives from rubber to infant formulas is quite great. In the U.S.A., research by Havery and Fazio (17, 20) demonstrated that 5 to 38% of the nitrosamines in rubber nipples migrated to liquid infant formulas or milk during sterilization (Table IV). They also found that not all the nitrosamines leached out during a single sterilization process. These compounds tended to leach out continuously, although in progressively decreasing amounts, on repeated sterilization.

Sen et al. (19) also reported similar results except that they noted migration of only 1-5% of nitrosamines from rubber nipples to infant formulas or orange juice (Table IV). The differences observed in the two studies (17, 19, and 20) could be attributed to the fact that in the study of Sen et al. (19) the nipples were not sterilized with the infant foods but were shaken with infant formulas or orange juice for 1 hr at 40°C.

It has been suggested that additional amounts of nitrosamines might be formed in babies' stomachs due to the interaction of salivary nitrite with ingested amine additives leached out from rubber nipples. Spiegelhalder and Preussmann (21) reported that 10-

Table IV. Migration of Nitrosamines from Rubber Nipples to Infant Formula, Milk, and Orange Juice

Experiment	Nipple used	Levels of nitrosamines in nipples (ppb)				% Migration into infant foods				Reference
		NDMA	NDEA	NDBA	NPIP	NDMA	NDEA	NDBA	NPIP	
Nipple sterilized in bottle with milk	(a)	64	28			17	20			(17)
	(b)		57		281		20		8	(17)
	(c)	2	11	2		18	5	29		(20)
Nipple sterilized in bottle with infant formula	(d)		29		165		18		5	(20)
	(e)		12	47			17	4		(20)
	(f)		27		114		19		6	(20)
Nipple shaken with infant formula	(g)	4	58	1030	459	4	3.5		1.1	(19)
	(h)	5	83	2052	180	9	1.4	5	4.5	(19)
	(i)	204		15		1.4		3.6		(19)
	(j)							N^{a}		(19)
Nipple shaken with orange juice	(g)	4	58	1030	459	N	1.9		N	(19)
	(h)	5	83	2052	180	N	1.5	2	3.8	(19)
	(i)							2.5		(19)

a N = none detected

(Adapted from Ref. 17, 19, 20.)

22,000 ppb (expressed in terms of the weight of a nipple) of certain
nitrosamines were formed when rubber nipples were incubated with
artificial saliva under simulated gastric conditions. Since the
nitrosation rate of a secondary amine is inversely proportional to
its basicity (11), the most weakly basic amines like methylphenyl-
amine and ethyphenylamine were found to produce the largest amounts
of nitrosamines under the above conditions. These findings led the
German government to pass a regulation that put a limit of 200 ppb
total leachable amines from baby bottle rubber nipples as determined
by the above method (21).

Recent studies by Sen et al. (22) suggest that in vivo nitrosation is unlikely to occur in the presence of infant formulas or fruit
juices. All infant formulas and fruit juices tested markedly
inhibited (up to 99%) the formation of nitrosamines, including that
from the weakly basic ethylphenylamine. Cow's milk was somewhat less
efficient in this respect. The presence of various N-nitrosation
inhibitors such as vitamin C, ascorbyl palmitate, α-tocopherol, and
naturally occurring phenols (e.g., chlorogenic acid, phloridzin) in
these products (12, 13, 22-24) was believed to be responsible for
this observed inhibition. A summary of the findings (22) is
presented in Table V.

As was observed by Havery and Fazio (17) for nitrosamines, Sen
et al. (22) also noted that amines in rubber nipples continued to
migrate into liquid infant foods even after repeated sterilization
(nipple boiled in water for 5 min for each sterilization). For
example, enough amines migrated into artificial saliva from a nipple,
which had already been sterilized six times, to yield up to 2,200 ppb
NEPhA and 226 ppb NDMA when tested by the German artificial saliva
method (21). In the presence of orange juice or infant formula,
however, nitrosamine formation was markedly inhibited. Therefore,
the health hazard, if any, that could arise due to the formation of
nitrosamines in babies' stomachs from ingested amines, might not be
as great as previously thought because of the modulating effect of
various food ingredients.

It should be mentioned that recent survey data from both the USA
(25) and Canada (22, 26) suggest a significant reduction in the
levels of both nitrosamines and amine precursors in baby bottle
rubber nipples from that observed previously. It is believed that
this has been achieved by modifying rubber curing formulations,
probably involving the use of nonamine accelerators or of amines that
do not form carcinogenic nitrosamines.

Other rubber products that may come in contact with foods
include rubber gloves used during handling of foods, rubber hoses and
tubings used in food processing industries or in milking machines
(18), and elastic rubber nettings used for packaging both cured and
uncured meats. Although traces of certain volatile nitrosamines and
diphenylamine have been detected, respectively, in rubber gloves and
water flowing through rubber tubing (15, 18), no data are available
with regard to the migration of these chemicals to foods. Further
research might be desirable in this area.

Elastic rubber nettings are often used for packaging cured pork
products (e.g., ham, cottage rolls, pork picnic shoulders, sweet
pickled pork cottage rolls) as well as some uncured meats (e.g.,
roast beef). They are used mainly for holding the meat pieces

Table V. Inhibition of Nitrosamine Formation by Various Liquid Infant Foods under Simulated Gastric Conditions

Nipple used	Nitrosamine detected	Percent inhibition[a] of nitrosamine formation from amines leached out from nipple				
		cow's milk	milk-based formula[b]	soy-based formula[b]	orange juice	apple juice
brand A	NDMA	96-100	97-100	100	100	100
	NDBA	26-63	65-99	100	94-100	100
	NPIP	30-90	100	100	100	100
	NEPhA	15-37	99-100	97	94-99	99
brand B	NDEA	not tested	97	100	99-100	100
	NPIP		91	100	94-100	100

a Compared to that formed in the absence of food, i.e., in artificial saliva.
b Contains ascorbyl palmitate and α-tocopherol as additives.
(Source: Reproduced from ref. 22. Copyright 1985 American Chemical Society.)

together during processing and cooking. Consumers usually keep the nettings on the products during warming or cooking (for raw products) at home. Sen et al. (27) recently reported finding NDEA and NDBA in rubber nettings. Analysis of cured meats packaged in such nettings indicated the presence of higher levels of nitrosamines than could be accounted for from that originally found in the unused nettings. This suggested formation of additional amounts of nitrosamines due to the interaction of nitrite, used in the meat curing process, with migrating amine derivatives (e.g., dibutyldithiocarbamate) from the rubber netting. When unused nettings were incubated with nitrite under mildly acidic conditions, excessively high levels of the same nitrosamines were formed suggesting the presence of the corresponding amine precursors in the nettings. No nitrosamines were found in similar cured pork products packaged in cotton nettings or plastic wrappings. The researchers concluded that the use of rubber nettings was the main reason for the occurrence of NDEA and NDBA (the predominant nitrosamine found) in the cured meats. These workers did not extend their studies to uncured meats because the chance of formation of nitrosamines in such cases would be very remote, due to lack of nitrite.

Sen et al. (27) also determined the concentration of NDEA and NDBA in different cross sections of cured pork products packaged in rubber nettings as well as that in the corresponding used nettings. This allowed the determination of the extent of the migration of the nitrosamines. The used nettings always contained the highest levels, followed by gradually decreasing levels in the meat's outermost ~5 mm layer, in the second ~5 mm layer, and practically none in the meat taken from the center. These findings strongly supported the conclusion that cured meats wrapped in rubber netting may contain nitrosamines due to migration of nitrosamines or amine precursors from the netting. Table VI presents three typical examples from their studies.

Further collaborative research with industry is continuing in the author's laboratory. It is hoped that, as in the case of baby bottle rubber nipples, the rubber netting industry will soon reformulate its rubber curing process that will eventually minimize or eliminate nitrosamine formation in such cured meat products.

Conclusion

From the limited information available to date, it appears that migration or formation of nitrosamines from food contact materials has mainly been a problem with rubber-based products. Although fairly high levels of MOR have been shown to migrate to foods from wax-based or paper-based food contact materials, conclusive evidence of formation of NMOR (either in vivo or in vitro) from the migrated MOR has been lacking. Further research on this or other unexamined food contact materials is desirable.

Table VI. Volatile Nitrosamine Contents of Nettings and Cured Pork Products Packaged in Rubber or Cotton Nettings

portion analyzed	sweet pickled pork cottage roll (rubber netting) NDBA (ppb)	smoked ham[a] (rubber netting) NDEA (ppb)	smoked ham[a] (rubber netting) NDBA (ppb)	smoked ham[a] (cotton netting) NDEA (ppb)	smoked ham[a] (cotton netting) NDBA (ppb)
unused netting	4.6	trace[b]	trace	N[c]	N
used netting	504	4.9	104	N	N
meat from outermost ~5 mm layer	57	6.9	33.3	N	N
meat from second ~5 mm layer	11	4.9	5.3	not analyzed	
meat from center	2.2	0.8	N	N	N
whole homogenized meat	19.5	2.4	5.9	N	N

a Also contained traces of N-nitrosothiazolidine which is produced as a result of smoking and not produced from rubber or cotton nettings.

b trace = <1 ppb.

c N = None detected (detection limit, 0.1-1 ppb).

(Source: Data taken from ref. 27. Copyright 1987 American Chemical Society.)

Literature Cited

1. Preussmann, R.; Eisenbrand, G. In Chemical Carcinogens; Searle, C.E., Ed.; ACS Monograph 182; American Chemical Society: Washington, D.C., 1984; 2nd edition, p 829.
2. Gray, J.I. In N-Nitroso Compounds; Scanlan, R.A.; Tannenbaum, S.R., Eds.; ACS Symposium Series 174; American Chemical Society: Washington, D.C., 1981; p 165.
3. Sen, N.P.; In Diet Nutrition and Cancer: A Critical Evaluation; Reddy, B.S.; Cohen, L.A., Eds.; CRC Press, Inc.: Boca Raton, Florida, 1986; p 135.
4. Hoffmann, D.; Brunnemann, K.D.; Adams, J.D.; Rivenson, A.; Hecht, S.S. In Nitrosamines and Human Cancer; Magee, P.N., Ed.; Banbury Report No. 12; Cold Spring Harbor Laboratory: N.Y., 1982; p 211.
5. Hotchkiss, J.H.; Vecchio, A.J. J. Food Sci. 1983, 48, 240-242.
6. Spiegelhalder, B.; Eisenbrand, G.; Preussmann, R. Angew. Chem. Int. Ed. Engl. 1978, 17, 367.
7. Sen, N.P.; Baddoo, P.A. J. Food Sci. 1986, 51, 216-217.
8. Kester, J.J.; Fennema, O.R. Food Technol. 1986, 40(12), 47-59.
9. The Merck Index, 9th edition; Merck & Co., Inc.: Rahway, N.J.; 1976, p 815.
10. Sen, N.P.; Baddoo, P.A. Proc. 47th Ann. Mtg. Food Technologists, June 16-19, 1987, Las Vegas, N.E.; Abstr. No. 370.
11. Mirvish, S.S. Toxicol. & App. Pharmacol. 1975, 31, 325-351.
12. Wilson, E.L. J. Sci. Food & Agric. 1981, 32, 257-264.
13. Stavric, B.; Klassen, R.; Matula, T. Proc. 193rd Am. Chem. Soc. Natl. Mtg.; April 5-10, 1987, Denver, Co.; Abstr. No. AGFD 45.
14. Fajen, J.M.; Carson, G.A.; Rounbehler, D.P.; Fan, T.Y.; Vita, R.; Goff, U.E.; Wolf, M.H.; Edwards, G.S.; Fine, D.H.; Reinhold, V.; Biemann, K. Science, 1979, 205, 1262-1264.
15. Ireland, C.B.; Hytrek, F.P.; Lasoski, B.A. Am. Ind. Hyg. Assoc. J. 1980, 41, 895-900.
16. Spiegelhalder, B.; Preussmann, R. Carcinogenesis, 1983, 4, 1147-1152.
17. Havery, D.C.; Fazio, T. Food Chem. Toxicol. 1982, 20, 939-944.
18. Babish, J.G.; Hotchkiss, J.H.; Wachs, T.; Vecchio, A.J.; Gutenmann, H.; Lisk, D.J. J. Toxicol. Environ. Hlth. 1983, 11, 167-177.
19. Sen, N.P.; Seaman, S.W.; Clarkson, S.G.; Garrod, F.; Lalonde, P. IARC Sci. Publ. 1984, 57, 51-57.
20. Havery, D.C.; Fazio, T. J. Assoc. Offic. Anal. Chem. 1983, 66, 1500-1503.
21. Spiegelhalder, B.; Preussmann, R. IARC Sci. Publ. 1982, 41, 231-243.
22. Sen, N.P.; Kushwaha, S.C.; Seaman, S.W.; Clarkson, S.G. J. Agric. Food Chem. 1985, 33, 428-433.
23. Newmark, H.L.; Mergens, W.J. In Gastrointestinal Cancer: Endogenous Factors; Bruce, S.R.; Correa, P.; Lipkin, M.; Tannenbaum, S.R.; Wilkins, T., Eds.; Banbury Report No. 7; Cold Spring Harbor Laboratory: N.Y., 1981; p 285.
24. Stich, H.F.; Rosin, M.P. In Nutritional and Toxicological Aspects of Food Safety; Friedman, M., Ed.; Plenum Press: New York, N.Y., 1984; p 1.

25. Havery, D.C.; Perfetti, G.A.; Canas, B.J.; Fazio, T. Food Chem. Toxicol. 1985, 23, 991-993.
26. Kushwaha, S.C.; Sen, N.P. Proc. 1987 AOAC Spring Training Workshop, April 27-30, 1987, Ottawa, Canada.
27. Sen, N.P.; Baddoo, P.A.; Seaman, S.W. J. Agric. Food Chem. 1987, 35, 346-350.

RECEIVED October 30, 1987

Chapter 13

Migration of Packaging Components to Foods

Regulatory Considerations

C. V. Breder

Keller and Heckman, 1150 17th Street, NW, Washington, DC 20036

The migration of packaging components to foods involves scientific and regulatory considerations. As a staff scientist at a law firm specializing in a wide range of regulatory issues, I am involved in a process where we routinely advise clients regarding the FDA status of their materials, recommend appropriate testing, review test results, prepare food additive petitions, provide legal opinions, etc. This paper will address the regulatory considerations by discussing the definition of a food additive, describing our firm's approach to establishing the FDA status of any food-contact material, and, finally, illustrating the discussion by means of a specific hypothetical example. The attorneys with whom I work have thoroughly reviewed this manuscript, and although I hope it offers helpful technical and scientific guidance, it should not be considered advice about the law concerning your specific food contact applications. A more exhaustive discussion of the pertinent legal concepts regarding the FDA compliance of food packaging materials can be found elsewhere.(1)

BACKGROUND

To provide a basis for our discussion of the regulatory considerations concerning the migration of packaging components to food, let me briefly review the applicable legal definitions and concepts surrounding the regulation of food packaging materials. We begin with Section 409 of the Federal Food, Drug and Cosmetic Act (Act) which prohibits the use of a "food additive" unless the additive is used in conformity with an FDA regulation. The term "food additive" is defined in relevant part in Section 201(s) of the Act as:

> [A]ny substance the intended use of which results or may reasonably be expected to result, directly or

0097–6156/88/0365–0159$06.00/0
© 1988 American Chemical Society

indirectly, in its becoming a component . . . of any
food . . ., if such substance is not generally recognized, among experts qualified by scientific training
and experience . . . to be safe under the conditions
of its intended use; except that such term does not
include . . .
(4) any substance used in accordance with a sanction
or approval granted prior to the enactment of this
paragraph. . . .

The term "food additive" plainly includes not only substances that are intentionally added to foods, such direct food
additives as antioxidants, and flavors, but also substances that
are not intentionally added but nevertheless contact and are
reasonably expected to migrate to food. These are the so-called
indirect food additives. As the title of my presentation suggests, I will devote the bulk of my discussion to indirect food
additives.

But just as, according to the cliche, all that glitters is
not gold, all materials contacting food are not food additives.
Under the above definition three categories of substances that
come into contact with food are not "food additives" and are not
subject to FDA regulation. They are substances that are:
     (1) prior-sanctioned
     (2) generally recognized as safe (GRAS); or
     (3) not reasonably expected to become a "component" of
          food

"Prior-sanctioned" means that the substance can be used in
accordance with a sanction or approval granted prior to the
enactment of the 1958 Food Additive Amendment.

Many other food contact materials enjoyed a long history of
safe use prior to the enactment of the 1958 Food Additive
Amendment. These materials were thus deemed to be generally
recognized as safe (GRAS) and it was felt that it would be ridi-
culous to deal with them further. Since 1958, of course,
additional substances have been affirmed as GRAS by experts
qualified by scientific training and experience.

The last category involves an important concept --"not
reasonably expected to become a component of food". Obviously,
consideration of this concept requires some feel for the
necessary analytical method sensitivity required to determine if
a food contact substance is reasonably expected to become a
"component" of food.

In 1958 when the Food Additive Amendment was enacted,
analytical methods were generally capable of reliably determining substances at concentrations of a few parts per million
(ppm). At that time, therefore, 1 ppm was a reasonable sensitivity to use when determining if a substance should be considered a "component" of food.

But while the law remained unchanged, analytical chemistry
made great advances. The ability to detect increasingly small
levels raised more and more questions about what is considered
"reasonably expected" to become a food additive. Some important

guidance on this question was provided by a prominent FDA
staffer named Les Ramsey in a 1969 speech before the National
Technical Conference of the Society of Plastics Engineers. He
proposed that the regulations be amended to permit, among other
things, components of food-contact articles provided any substance so used contribute no more than 0.05 ppm (50 ppb) to the
contacted food.   This proposal has come to be known as the
"Ramsey" proposal.   Although it was never formally adopted, the
principles are still considered valid.

Since 1969 we have relied on the "Ramsey" proposal in
establishing the appropriate analytical method sensitivity to
determine if a food contact substance is reasonably expected to
become a component of food and therefore, a food additive. Our
position is that the 50 ppb sensitivity can be conservatively
recommended in those cases where the food additive would not
pose some toxicological or public health concern, or, where the
use of the additive is not expected to lead to wide consumer
exposure such as in a carbonated beverage bottle.   In other
cases where there is some toxicological concern about the additive and/or it will lead to wide consumer exposure, we may
recommend a method sensitive to 10 ppb or less.   Of course,
carcinogens, heavy metals, pesticides and other substances of
high toxicological concern require separate and individual
study.

Thus, to summarize what we mean by not reasonably expected
to become a "component of food", if a food contact substance is
non-detected in extracts of appropriately conducted extraction
studies using an adequate analytical method sensitivity, it is
our opinion the substance is not expected to become a "food
additive" requiring preclearance or regulation by FDA.

If, however, a food contact substance does become a component of food and is not GRAS or prior-sanctioned, its use must be
authorized by food additive regulations promulgated by FDA.

DISCUSSION

There is a logical approach to establishing the FDA status of
any given food-contact material.   My concern in presenting this
approach is that it is easy to assume that every situation can
be evaluated simply when, in fact, each situation is unique and
full of its own themes and variations, each of which may require
separate and individual study.   So, please be aware that this
discussion may appear simpler than reality justifies. Nevertheless, the principles involved are certainly valid for
establishing the FDA status for any food-contact material.

With this caveat in mind, what is this logical approach to
establishing the FDA status of a food contact material?

1.   Determine the Identity of the Food Contact Substance(s), Possible Migrant(s) and the Intended Conditions of Use. First
and most importantly, the identity of the food-contact substance
and its intended conditions of use must be accurately deter-

mined. This determination will serve as the basis for the rest
of the evaluation process. For example, is the potential food
additive a polymer, monomer, oligomer, adjuvant, etc.? Will the
material be used in bottles, film, laminates or pouches? How
thick will these products be? Will the material be exposed to
high temperatures, room temperature, or only refrigerator temperatures? What types of foods will contact the material--aque-
ous, fatty, acidic, or alcoholic? Is the material to be used in
a single service or repeated use application? Will it be separated from food contact by a functional barrier? The answers to
these, and possibly other, questions must be known before
proceeding to the next step of the evaluation.

**2. Research the Existence of Applicable Clearances Under
Existing Food Additive Regulations, Prior-Sanction or GRAS Listings.** Having identified the potential food additive(s) and the
intended conditions of use, a diligent effort must be made to
determine if these materials are cleared by existing regulations, are GRAS or prior-sanctioned. Believe me, this is the
quickest and least expensive way to establish the FDA status of
food-contact materials because if appropriate clearances can be
found, no further work is necessary.

**3. By Means of a Worst-Case Calculation Assuming 100% Migra-
tion, Determine if the Material is Expected to Become a Food
Additive.** Thirdly, if no applicable clearance can be found, from
the information obtained in the first step it may be possible in
some situations to assume total migration to the food and calculate the maximum potential concentration of the additive in the
food. If this calculated concentration would be non-detected
using the appropriate analytical method sensitivity, and, it
poses no undue safety concerns, one may be able to conclude that
the component is not a food additive and can be used without
further work or consultation with FDA. A large comfort factor
is provided in this approach because the level of migration is
almost always far less than 100%.

To give an example, if a 1/8th inch thick food conveyor
belt containing 20 parts per million (ppm) of an antioxidant is
in contact with 100 lbs. food/in^2 over the lifetime of the belt,
it can be calculated that the resultant concentration of the
antioxidant in the food is approximately 1 part per billion (1
ppb) assuming it all migrates. Clearly this level is vanish-
ingly small and would be non-detected in the food. Our view,
therefore, is that it is not a food additive. In this case, you
may properly conclude that the antioxidant is not expected to
become a component of food and it can be used as intended without consulting FDA.

**4. Conduct Appropriate Extraction Studies to Determine if and
to What Extent Migration Occurs.** If step 3 does not work out,
as is often the case, appropriate extraction studies must be
designed which will simulate and, to some extent, exaggerate the

intended conditions of use. Suppose, however, the intended conditions of use could not be accurately determined in the first step, then one can assume the worst-case application-- i.e., thickest product, highest temperature, most extractive foods, etc. and proceed with extraction studies on this basis. All other applications for the material would be automatically covered because they would lead to less migration. If it doesn't work out, the end-use applications for the material will have to be limited depending on the extraction data and supporting toxicity data.

Depending upon the intended conditions of use, or the worst-case application as just discussed, the proper food simu- lating solvents, sample surface area/solvent volume ratio, temp- eratures, times and analytical method sensitivity must all be determined and used appropriately. For example, the traditional aqueous food simulating solvents are: water for aqueous foods, 3% acetic acid for acidic foods, 8% and/or 50% ethanol to simu- late alcoholic foods. (Incidentally, FDA is considering the use of only 8% ethanol in place of water, 3% acetic acid and 8% ethanol for some applications).

Heptane has traditionally been used to simulate fatty foods. It is widely recognized, however, that heptane exagger- ates the extraction potential of fatty foods from many food- contact surfaces. For this reason, FDA has permited the heptane extractives to be divided by 5 to offset this distortion. In an effort to better simulate fatty foods, FDA is now considering abandoning the use of heptane, and requiring the use of various water/alcohol mixtures or an actual fat such as corn oil. For example, 95% ethanol is now recommended as the fatty food simu- lant for polyolefins and 50% ethanol is recommended for poly- styrene and polyethylene terephthalate. FDA is still looking for appropriate water/ethanol mixtures that can be used as fatty food simulants for other polymer systems.

For general packaging applications, FDA assumes 10 grams of food will contact each square inch of package surface area. Therefore, the extractives found, if any, should be expressed as the concentration in food based on 10 g/in^2. Of course, for repeated-use applications, such as the conveyor belt example discussed earlier in which a large volume of food contacts the material over a period of time, a larger volume to surface area ratio may be used.

The times and temperatures employed in the extraction study will vary widely depending upon the intended conditions of use. For example, a can coating will have to be extracted with water at 250°F for 2 hours followed by 10 days storage at 120°F to simulate cooking or sterilizing conditions and subsequent room temperature shelf storage of aqueous foods in the can. On the other hand, a soft drink container need only be extracted with 3% acetic acid at 120°F for 30 days to simulate room temp- erature shelf storage. This is why it is critical to know how a material is to be used. Not knowing the intended conditions of use may lead one to greatly under- or over-estimate the amount of migration that may occur in commercial use.

Finally, spiking and recovery studies must be carried out to validate any analytical methods used. This work is critical to establishing the credibility of the results obtained.

Of course, if no migration is found, the component is not a "food additive" and can be used as intended without consultation with or permission by FDA. If the component is found in the extracts, it is an uncleared food additive requiring an FDA regulation.

5.   If There is Migration, Calculate an Estimated Dietary Concentration of the Migrant(s) in the Total Diet. Armed with the results of the extraction studies and assuming that migration was found, the amount of the food additive that may become a component of the daily diet must be calculated. FDA has procedures for making these calculations. Basically, they involve a consideration of the type of food contact material-- e.g., polymer coated metal, polymer coated paper, polyolefins -- and, the types of foods these materials contact--i.e., aqueous, acidic, alcoholic, or fatty.

FDA has determined the food-type distribution for most food contact materials. For example, of the foods that contact polyolefins, 67% are aqueous, 1% are acidic, 1% are alcoholic and 31% are fatty. Of course, other polymers have different food-type distribution factors. Thus, depending upon the particular food contact applications, the extraction results using the appropriate food simulating solvents can be converted to an estimate of the amount migrating to packaged food.

FDA has also developed information that reflects the percentage of the daily diet which contacts most types of packaging materials. For example, polymer coated metal contacts approximately 17% of the diet, polymer coated paper--21%, and polyolefins--33%. These food contact percentages are called the consumption factor, or CF. Thus, the calculated concentration of the component in the packaged food can be converted to a concentration in the total diet by multiplying by the consumption factor for a given food contact material.

In summary, the concentration of the additive in the total diet is obtained by multiplying the extraction results in the various food simulating solvents by the appropriate food-type distribution factors and subsequently by the appropriate consumption factor for the food contact material. A hypothetical situation is shown in Table I.

6.   Obtain Appropriate Toxicology Data. Knowing the estimated concentration of the additive in the diet permits one to determine what toxicology data are required to support the safe uses of the food additive. Again, FDA requires different toxicology tests for indirect additives based upon the dietary concentration. For example, studies based on the calculated dietary concentration are generally required (see Table II.)

In our extraction study example, we obtained a calculated dietary concentration of 0.4 ppm. This dietary level would require two 90-day feeding studies.

Table I. Concentration of additives in total diet

Polymer	CF	Food-Type Distribution			
		Aqueous	Acidic	Alcoholic	Fatty
Polyolefins	0.328	0.67	0.01	0.01	0.31

Hypothetical Migration Results: aqueous   :   0.2 ppm
                                 acidic   : 10.3 ppm
                                 alcoholic :   0.7 ppm
                                 fatty   :   3.1 ppm

Concentration of additive in contacted food, $<m>$:
$$<m> = (0.67)(0.2) + (0.01)(10.3) + (0.01)(0.7) + (0.31)(3.1)$$
$$= 1.2 \text{ ppm}$$

Concentration of additive in the total diet:
$$CF \times <m> = 0.328 \times 1.2$$
$$= 0.4 \text{ ppm}$$

Table II. Concentration of indirect additives in total diet

Concentration of Additive in Diet	Required Toxicology Studies	Approximate Costs
1. 0.05 ppm or less	acute oral toxicity	$5,000
2. 0.05 - 1 ppm	two 90-day sub-chronic studies (one in a non rodent)	$125,000 each
3. 1 ppm or greater	2-year chronic study in rodent and non-rodent	$1,250,000

Generally, only an acute oral toxicity study is required for additives present in the diet at a concentration of 50 ppb or less. However, for higher dietary concentrations, the required sub-chronic or chronic studies should be conducted in such a way as to provide a no-observable-effect level, the so-called NOEL for the chemical. FDA takes the NOEL from a two-year feeding study in the most sensitive species and divides it by a safety factor of 100 to obtain the acceptable daily intake (ADI) for the additive. If two 90-day feeding studies were used, FDA takes the lowest NOEL obtained in these studies and divides by a safety factor of 1,000 to obtain the ADI. The chemical additive is deemed safe if the ADI is greater than the calculated dietary concentration.

7.   Consider GRAS Position. If Necessary, File a Food Additive Petition and Obtain Regulation. Finally, if it is determined by appropriate extraction tests that you have an uncleared food additive, in some cases you might on the basis of the amount of migration observed and the appropriate toxicology data be able to conclude that the intended uses of the food contact substance(s) are GRAS. If this is not a reasonable position to take (a subject beyond the scope of this paper), then you must file a food additive petition seeking a regulation for the intended use of the additive.

Briefly, the petition must contain:
A.   Identification of the additive
B.   Amount of additive proposed for use
C.   Intended technical effect
D.   Method for determining presence of additive in food
E.   Safety of the additive
F.   Proposed tolerances
G.   Modification of existing regulation
H.   Environmental Assessment

HYPOTHETICAL EXAMPLE

Consider a hypothetical situation in which you have discovered a chemical substance that, when blended into polypropylene at a level of 0.1%, is effective in preventing the discoloration of ketchup packaged in polypropylene bottles. The chemical is not intended to become a component of ketchup to be effective. You have determined, properly but laboriously, that the chemical is not cleared in any specific FDA regulation for contact with food nor does it appear in any GRAS or prior-sanction listings. By assuming 100% migration of the chemical from the bottle to the ketchup, you have calculated a worst-case concentration of approximately 30 ppm. Clearly, the chemical at such a concentration would be considered an uncleared food additive.

With this background information in mind, the question becomes, what is the FDA status of the chemical and how is that determined? Let's look at the various FDA status possibilities for this intended use. These possibilities are:

a. Chemical does <u>not</u> migrate to food -- no "food additive" and can be used without FDA regulation.

b. Chemical migrates to food -- must file a food additive petition and obtain FDA regulation.

How do we know which of these scenarios is operational under the intended conditions of use? Appropriate extraction studies must be carried out to determine if the chemical is expected to become a component of food, i.e., a "food additive", under the intended conditions of use. To simulate, and to exaggerate to some extent, how the product will be used, the right simulating solvents, correct solvent volume to surface area ratio, temperatures, storage times, and analytical methods must be chosen.

For the intended conditions of use anticipated for this application, the following protocol is suggested:

Solvent — 3% acetic acid

Solvent volume/
 surface area — same as in smallest ketchup bottle

Temperature — Hot fill and let cool to 120°F

Time — 10 days at 120°F

Analytical Method — 50 parts per billion (ppb) sensitivity

Three percent acetic acid is used to simulate the acidic food ketchup. None of the other commonly used food simulating solvents, such as water, 8 or 50% alcohol or heptane are appropriate in this case.

To provide a worst-case exposure, the volume of solvent in contact with the polypropylene should be the same volume/surface area ratio that is provided by the smallest ketchup bottle expected to be used. By selecting the smallest ketchup bottle, we are covering the worst-case volume-to-surface area anticipated under the intended conditions of use. Thus, whatever migration, if any, found at this volume/surface area ratio will represent the highest level anticipated knowing that larger containers will lead to lower levels.

To simulate the actual intended use during the filling operation, the temperature of the solvent that is added to the bottle should be the same as, or to represent mild exaggeration, a few degrees higher than the temperature at which ketchup is bottled. Following the filling of the bottle, the contents should be allowed to cool to 120°F and maintained at this temperature for 10 days. FDA considers that 120°F exaggerates room temperature storage. Thus, by maintaining the extraction test at this temperature for 10 days, the extended storage at room temperature encountered by the ketchup in commerce is adequately simulated.

Since the chemical in question is not of particular toxic concern and will not be used in a high consumer exposure application such as in a carbonated beverage bottle, 50 ppb is the appropriate analytical method sensitivity.

If the aforementioned extraction study reveals that the chemical is non-detected in the extracts, then, in my opinion, you can conclude the chemical is not a "food additive" and that its use is in compliance with the Federal Food, Drug and Cosmetic Act and all applicable Food Additive Regulations. Thus, there is no regulatory or legal need to discuss the matter with FDA, ask its opinion, or file a food additive petition.

On the other hand, if the chemical is found to migrate, it is a food additive and can only be used if it is cleared by a new FDA regulation. Such a regulation is obtained by filing a food additive petition.

DIRECT FOOD ADDITIVES

The discussions and hypothetical example have focused up to this point on indirect additives. However, when packaging materials are used to intentionally convey specific additives to the contained food, the additives would be considered by FDA to be direct food additives. Although the same consumer exposure and toxicity considerations for indirect food additives generally apply to direct food additives, there is an additional consideration to be aware of -- a standardized food may be involved. Many foods are "standard foods" and must meet their respective standards of identity before they can be properly labeled.

For example, had the previous hypothetical example involved the direct addition of the discoloration inhibiting chemical to the ketchup, we would be adding a new substance to a standardized food. Ketchup is a standard food which is defined by its standard of identity in 21 C.F.R. Sec. 155.194. The only optional ingredients permitted in ketchup are:

(i)   vinegars
(ii)   nutritive carbohydrate sweeteners
(iii)   spices, flavoring, onions, or garlic

Thus, if the chemical which is intended to be added to ketchup is not permitted by any of these three categories, the resulting ketchup is not in compliance with its standard of identity and cannot be labeled as such. Under 21 C.F.R. Sec. 130.8(a), a food does not conform to its standard of identity if

"it contains an ingredient for which no provision is made in such definition and standard, unless such ingredient is an incidental additive introduced at a non-functional and insignificant level. . . ."

Thus, the addition of any new functional ingredient to ketchup will not be permitted without the submission of a petition to amend the food standard. On the other hand, as given in Section 130.8(a), the incidental migration of the chemical into ketchup, as discussed in our first hypothetical example, is permitted.

In the present case, therefore, two petitions are required--one is a food additive petition to clear the substance from a safety standpoint and the other is a petition to amend the standard of identity for ketchup to include the deliberate addition of the chemical additive. It should be noted that the second petition is not to be taken lightly because there must be general industry agreement to amend the standard before it can be amended. If, for example, FDA receives a number of adverse industry comments to its proposal to amend the ketchup standard, then it may choose not to amend it. If this happens, there will be no market for the chemical when used as intended in our direct additive example.

SUMMARY

Hopefully, the discussions and hypothetical examples will heighten your awareness about the regulatory requirements relevant to the migration of packaging components to food and reinforce your understanding of our approach to establishing the FDA status of a food contact material.

In summary, I reiterate the seven basic steps in establishing the FDA status of food-contact substances. They are: (1) determining the identity and intended conditions of use of the food contact substance(s), (2) assessing if these substances are already cleared, (3) calculating, if possible, if the substance can be considered not to be a "food additive" under the conditions of use, (4) conducting appropriate extraction studies, (5) calculating the estimated dietary concentration of the additive, (6) obtaining adequate toxicology data to support the safety of the intended use, and (7) if necessary, filing a food additive petition.

Reference

1. Heckman, Jerome H., *Food Drug Cosmetic Law Journal* 1987, 42, 38-39.

RECEIVED October 16, 1987

Chapter 14

Development of a Standard Test for Volatile Migrants from Polyester Trays

Sara J. Risch and Gary A. Reineccius

Department of Food Science and Nutrition, University of Minnesota, St. Paul, MN 55108

A test was developed to determine the concentration of
volatile migrants from thermoset polyester trays.
Existing FDA regulations required only testing for
total nonvolatile extractives. A test was needed to
measure the amount of volatile compounds migrating
from the trays into food products when used in a conventional oven. Initial testing indicated that both
vegetable and mineral oils suffered severe heat
induced degradation that would prevent isolation and
identification of compounds in the low parts per
billion. Silicone oil was tested for its suitability
as a fat simulant and found to be similar to vegetable oil in amount of oil absorbed and effect on surface hardness. The final test uses an emulsion with
76% water, 19% silicone oil and 5% gum arabic. The
emulsion is frozen on the trays, baked and an aliquot
taken for quantification by dynamic headspace
concentration/capillary gas chromatography.

In the past few years, we have seen an explosive growth in the
development of new packaging materials. There has been a push by
both food and plastic manufacturers and a pull by marketing personnel and consumers for better packages: more functional, attractive,
and convenient. One area in particular where convenience has been a
major factor is the trend in eating habits in the United States
toward more meals being eaten alone, increasing the demand for
single serving meals, often referred to as "TV dinners". A number
of food companies have upgraded these meals and created what they
refer to as gourmet dinners. The higher quality dinners also
required a package that presented a new and upscale image but also
had to meet the criterion of being dual-ovenable. One essential
part of this new concept was a package that presented a china-like
image. One word commonly used to describe the desired package or
plate is heft. The plate needed to be sturdy and appealing. The
industry's answer to these requirements was the thermoset polyester
tray. The tray functions well in the microwave oven where heating

0097–6156/88/0365–0170$06.00/0

times are in the range of six to eight minutes. In a conventional
oven, where the heating time is substantially longer (30-45 min) and
temperature higher (ca 350°F), the food baked on the trays
occassionally develops an off-taste or odor. This problem, as well
as concern about the regulatory status of the tray, led to the
development of an appropriate test to determine the levels of specific compounds migrating under intended use situations. The compounds of concern were those which were volatile and could be
monitored by gas chromatography.

   At the time this research was started, the existing Food and
Drug Administration (FDA) policy, as defined in Parts 170-189 in the
Code of Federal Regulations (CFR) (1) did not describe any regulations for this type of packaging material under the intended use
situation. As with most packaging materials, the specifications
were that the total amount of material (in units of mg/in^2)
migrating into extracting mediums could not exceed a certain level,
typically $0.5 mg/in^2$. Examples of the testing mediums are heptane
to simulate fatty foods, 3% acetic acid to simulate acidic foods,
and water for neutral foods. The times and temperatures of contacting for each medium were defined with the maximum temperature
generally $\leq$ 120°F. It was immediately obvious that none of these
testing media could be used to test for volatile migrants during
baking at 350°F for 30 minutes or more.

   Considerable information is available on related work being done
by member nations of the European Economic Community (EEC) to
establish uniform guidelines and regulations for packaging
materials. A review of the existing literature is not the intent of
this presentation, but it should be noted that Rossi (2,3) has
published two compilations of interlaboratory studies concerned with
migration of plastics materials into food simulants. Other work has
been concerned with both global or total and specific migration into
various edible oils and fatty foods (4,5,6).

   The first step was to find an appropriate food simulant. The
first suggestion was a food grade oil which would not evaporate
during heating. Several oils were tried, including Crisco (Procter
and Gamble), coconut, mineral, and castor oil. The first criterion
was that the oil had to be relatively free of volatile components so
that migrating compounds could be seen in a gas chromatographic (GC)
profile in the low parts per billion (ppb) concentration (the
detailed GC method of analysis will be described later). Crisco oil
met this criterion as it was received commercially (Figure 1A). The
other oils were initially contaminated and an attempt was made to
remove interferring volatile compounds by vacuum stripping in the
apparatus shown in Figure 2. The oil was introduced at a rate of
one drop per second into the tube filled with glass helices held at
55°C. The system was operated at an absolute pressure of ca. 2mm
Hg. The system was effective in removing volatile contaminants as
demonstrated in Figure 1B and C which contrast the GC profiles of
coconut fat before and after vacuum stripping. Of the 15 peaks with
area counts greater than 100, the average reduction in peak size
after stripping was 63% with a range from 0 to 95%.

   The second criterion was that the oil had to be stable to heat
at 350°F for at least 30 minutes. In this respect, none of the
vegetable oils was appropriate. All of them suffered severe heat

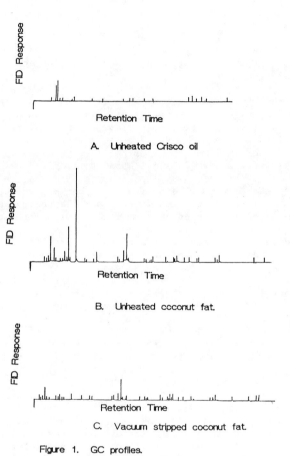

A. Unheated Crisco oil

B. Unheated coconut fat.

C. Vacuum stripped coconut fat.

Figure 1.   GC profiles.

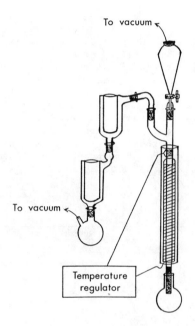

To vacuum

To vacuum

Temperature
regulator

Figure 2. Vacuum stripping apparatus.

induced degradation. A typical example is Crisco oil whose GC profiles after heating in glass and thermoset polyester trays are shown
in Figures 3 and 4, respectively. This is in contrast to the GC
profile of unheated Crisco, which was shown in Figure 1A.

The next oil tried was silicone oil which is not edible but is
stable at higher temperatures. Two tests, oil absorption and surface hardness, were used to demonstrate that silicone oil reacted
with the trays in a manner similar to vegetable oil. A method for
purifying the oil also had to be found.

Oil absorption was determined by drying ten identical trays at
250°F for one hour, cooling and recording the initial weight of
each. Five of the trays were filled with vegetable oil and five
with silicone oil. The filled dishes were subjected to five cycles
of heating at 250°F for one hour followed by cooling 20 minutes at
room temperature. After the final cooling, the oil was decanted,
trays wiped with cheesecloth, and then quickly wiped with a hexane
dampened cloth to remove any residual surface oil. The trays were
reweighed and the weight gain was taken as the amount of oil
absorbed. The results, listed in Table I, showed no significant
difference between the amount of vegetable oil absorbed and the
amount of silicone oil absorbed.

Table I. Oil absorption by polyester trays filled with either vegetable oil or silicone oil during five heating/cooling cycles

| Replicate | Weight gain in grams | |
	Vegetable Oil	Silicone Oil
1	0.03	0.02
2	0.04	0.03
3	0.03	0.01
4	0.01	0.03
5	0.08	0.04
Average	0.038	0.026
Standard deviation	0.026	0.011

The test of surface hardness was used to indicate whether silicone oil brought about changes in the surface of the trays that were
different than changes from contact with vegetable oil. Surface
hardness was measured in arbitrary units with a Barber-Colman surface hardness meter. Ten determinations of surface hardness were
made on each plate that had been used to determine oil absorption
for a total of 50 determinations on plates that had been heated with
vegetable oil and 50 on plates that had been heated with silicone
oil. The results (Table II) indicated no significant difference in
surface hardness between plates heated with the two types of oils.
Based on the results of the tests for oil absorption and surface
hardness, silicone oil was deemed to be a suitable fat simulant.

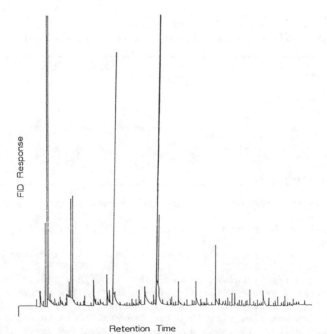

Retention Time

Figure 3. GC profile of Crisco oil baked in glass bowl.

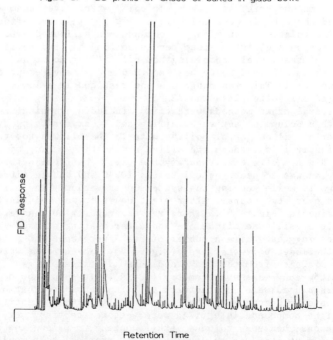

Retention Time

Figure 4. GC profile of Crisco oil baked in thermoset polyester tray.

Table II. Surface hardness of polyester trays filled with either
vegetable oil or silicone oil during five heating/cooling cycles

	Surface Hardness[a,b]	
Replicate	Vegetable Oil	Silicone Oil
1	44.3	44.6
2	45.1	44.2
3	45.9	46.2
4	45.1	44.7
5	40.5	40.2
Average[c]	45.2	44.9
Standard deviation	2.5	2.6

[a] Values for each replicate are the average of ten individual determinations.
[b] Values are expressed as arbitrary units of the surface hardness
meter.
[c] Average and standard deviation are based on all values obtained.

Silicone oil as received commercially is highly contaminated as is
demonstrated by the GC profile in Figure 5. The method of vacuum
stripping described earlier was ruled out for two reasons. First,
the method was slow with it taking several hours to purify 100 mL of
oil, and second, the method was not as effective for silicone oil as
it was for vegetable oils.

One method tried was steam stripping where live steam was
bubbled through silicone oil held at 150°C. This was effective in
removing volatile contaminants, but had one serious drawback. A
small percentage of the steam tended to condense and collect in the
bottom of the vessel containing hot oil. After about an hour, the
water that collected would boil and cause a massive eruption of
silicone oil. This was dangerous and resulted in a low yield of
purified oil, often less than 25%.

Several other methods were tried, including running the oil
through a column of activated silica and steam purifying by putting
a layer of oil on top of boiling water. The oil was too viscous to
efficiently flow through the silica and heating on top of boiling
water did not effectively purify the oil. The final method
attempted was to heat the oil to ca. 200°C and bubble nitrogen
through it while keeping the system under vacuum. The oil was
tested every two hours. After eight hours it was relatively free of
contaminants (Figure 6). There was essentially no change in the GC
profile of silicone oil before and after heating at 350°F for 30
minutes in glass while a number of peaks were apparent after heating
in a thermoset polyester tray (Figure 7). These peaks were presumed
to be migrants from the trays and were identified by gas
chromatography/mass spectrometry (GC/MS). It should be noted that
the trays used were manufactured to be known bad so that higher
amounts of migrants would be present to facilitate identification.
The eight compounds identified were acetone, tertiary butyl alcohol,
2-butanone, benzene, toluene, ethyl benzene, m-xylene and styrene.

The next step was to develop a test protocol that would
appropriately simulate the reheating of a frozen dinner in a conven-

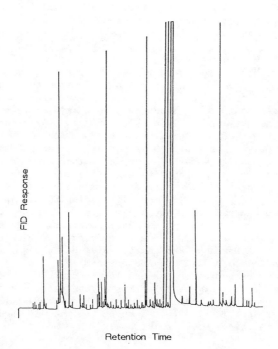

Figure 5.   GC profile of silicone oil as received commercially.

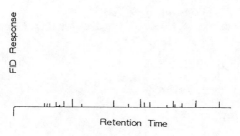

Figure 6.   GC profile of purified silicone oil.

tional oven. The use of pure silicone oil was ruled out because it
is unlikely that any dinner would be pure fat in contact with the
entire tray. An emulsion of silicone oil, water, and gum arabic was
formulated which could be put on a tray, frozen, and then baked.
The emulsion was made with 76 parts distilled water (Glenwood Co.,
Minneapolis, MN), 19 parts purified silicone oil, and 5 parts gum
arabic. A paste of gum arabic and a small amount of water was prepared in a mortar and pestle and then blended with the remaining
water and silicone oil at high speed for 1 minute in a Waring
blender. The prepared emulsion (150 mL) was poured into a tray, the
tray covered with foil, and frozen at -20°F for 24 hours. This
volume was conveniently held by the tray without overflowing. After
freezing, the foil was turned back ca. 1 inch along one edge (as
package directions for preparing a meal indicate), and the tray
baked in a preheated 375°F oven for 45 minutes. The longer cooking
time at a higher temperature than listed in the meal preparation
instructions represents an abuse situation so amounts of compounds
may be overestimated instead of the possibility of underestimating.
The test emulsion (ca. 120mL due to evaporation) was decanted from
the tray into a Pyrex beaker immediately upon removal from the oven
and allowed to cool 30 minutes before sampling. Some oil separation
did occur during baking so the sample was stirred to mix thoroughly
before a 15 mL aliquot was taken for GC analysis.

A Hewlett Packard model 7675A Purge and Trap sampler coupled to
a Hewlett Packard model 5840 gas chromatograph was used for analysis. The conditions were as follows:

```
Purge temperature:  Ambient
Purge flow:  50 mL/min.
Purge time:  20 min.
Purge gas:  Helium
Trap:  Tenax GC
Desorb time:  3 min.
Desorb temperature:  180°C
GC column:  30 m x 0.25 mm DB-5 (J & W Scientific)
Split:  50:1

Oven temperature profile:
    During Desorb:  -160°C, 3 min.
    Initial:  35°C, 4 min.
    Program:  8°C/min.
    Final:  200°C, 10 min.
```

The amount of each compound migrating from a tray into the
emulsion was quantitated by comparison to the peak areas obtained
when 10 ppb of benzene and 50 ppb of the other seven compounds identified were added to a sample of the emulsion and analyzed by the
same gas chromatographic procedure as outlined above. An example of
this is shown in Figure 8. From this standard run, a peak area per
ppb could be determined and the amount of compound calculated as
follows:

$$\frac{\text{Peak area sample}}{\text{Peak area per ppb in standard}} = \text{ppb in sample}$$

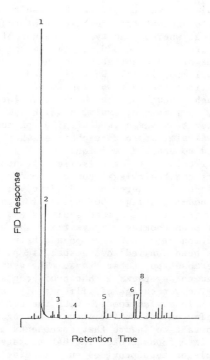

Retention Time

Figure 7.   GC profile of purified silicone oil baked
in thermoset polyester tray.

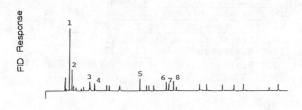

Retention Time

Figure 8.   GC profile of silicone oil : water emulsion
spiked with 50 ppb standard compounds.

One further calculation was to divide the result (ppb in sample) by 2. This takes into account that the amount of testing solution is approximately 120 g, whereas the typical weight of a dinner prepared on a tray ranges from 247 g to 320 g. The amount calculated is expressed in terms of a concentration so the estimated ppb likely to migrate into a full meal is half what migrated into the emulsion.

One final confirmation test was run to determine whether baking the emulsion brought about changes which would interfere with detection of migrants. A sample of emulsion which had been baked in a thermoset polyester tray was spiked with 50 ppb acetone and styrene. The GC response was similar to that obtained when 50 ppb acetone and styrene were added to unbaked emulsion.

The purpose of this test for volatile migrants was to permit tray manufacturers and their legal council to determine whether or not the tray could be considered a food additive. If the compounds were found to be non-migrating (<50 ppb or <10 ppb for known carcinogens), the tray is not a food additive and can be used as intended for dual oven cooking. Trays that have been tested have met the specification of volatile compounds being non-migrating.

This test has been applied to one specific package, single use dual-ovenable thermoset polyester trays. The method has the potential to be applied to any package that could contain the liquid testing medium. The percentage of silicone oil could be varied to simulate higher or lower fat foods. With the continued growth in the use of packaging materials for either cooking or reheating foods, it is important to insure that components are not migrating from the package into foods cooked in that package. The silicone oil and water emulsion test provides an effective means of testing high heat packaging materials under intended use situations, whether in a conventional or microwave oven.

Literature Cited
1. Code of Federal Regulations, Title 21. U.S. Government Printing Office, 1984.
2. Rossi, L. J. Assoc. Off. Anal. Chem. 1977, 60, 1282.
3. Rossi, L. J. Assoc. Off. Anal. Chem. 1981, 64, 697.
4. Frigge, K. Food Cosmet. Toxicol. 1972, 11, 815.
5. Freytag, W., Figge, K. and Bieber, W.-D. Deutsche Lebensmittel-Rundschau, 1984, 80, 333.
6. Gilbert, J., Startin, J.R. and McGuinness, J.D. Food Addit. Contam. 1986, 3, 133.

RECEIVED October 27, 1987

Chapter 15

Chemical Changes in Food Packaging Resulting from Ionizing Irradiation

Donald W. Thayer

Eastern Regional Research Center, U.S. Department of Agriculture, Agricultural Research Service, Philadelphia, PA 19118

Recent approvals of food irradiation processes
by the U.S. Food and Drug Administration have
led to a search for packaging approved for use
with ionizing radiation. Though 13 packaging
materials were approved several years ago as
food contactants for gamma irradiation up to
10 kGy at refrigeration temperatures and 4 packaging materials were approved for up to 60 kGy
at cryogenic temperatures, no currently used
packaging is approved for irradiated foods.
Extensive research was conducted by the
U.S. Army and others on the suitability of both
flexible packaging and metal cans for packaging
irradiated foods. The results of the studies of
packaging for irradiated foods will be described
and discussed in context of currently used packaging materials for non-irradiated meats and
poultry.

There have been several recent approvals for the use of ionizing
radiation treatments of food products and it is anticipated that
more will be forthcoming for the use of ionizing radiation
treatments to eliminate food-borne pathogens in (and incidentally
to increase shelf life of) meats and poultry. Though several
materials are approved as food contactants for gamma irradiation up
to a maximum dose of 10 kGy by the Food and Drug Administration,
the existing regulation (1) was written in the late 1960's to
provide packaging materials for the planned irradiation of bacon by
the U.S. Army. Only films were included in the regulation; no
multilayer materials were included. Furthermore, only gamma
radiation sources were included in the regulation. Today, stretch
or shrink packaging using bags or trays with overwrap is most
commonly used for meat and poultry. The packaging films usually
used for poultry have a high moisture barrier and a low oxygen
barrier. The packaging films used for red meat usually have both
high moisture and high oxygen barriers. Frequently the required

properties for packaging materials for both poultry and red meats
are best achieved using multilayer films. It is in the latter
that a problem may exist.
     The purpose of this manuscript is to review the extensive
research conducted by the U.S. Army and others on the suitability
of both flexible packaging and metal cans for packaging irradiated
foods. The results of the studies of packaging for irradiated
foods will be described and discussed in context of currently used
packaging materials for non-irradiated meats and poultry.

Packaging Materials Approved for Irradiated Foods

The regulation (1) states that the following materials may be
safely subjected to gamma radiation doses up to 10 kGy incidental
to the radiation treatment and processing of prepackaged foods:
1) wax-coated paperboard, 2) vegetable parchments, 3) nitrocellu-
lose-coated cellophane, 4) vinylidene chloride copolymer-coated
cellophane, 5) glassine paper, 6) polyolefin film, 7) polyethylene
film, 8) polyethylene terephthalate film, 9) polystyrene film, 10)
rubber hydrochloride film, 11) vinylidene chloride-vinyl chloride
copolymer film, 12) nylon 6 film, 13) nylon 11 film, and 14) vinyl
chloride-vinyl acetate copolymer film. In the regulation the
coated cellophanes are treated as a single class. The regulation
(1) also lists the approval of Kraft paper as a container for
flour with a dose not exceeding 0.5 kGy. Vegetable parchments,
polyethylene film, polyethylene terephthalate film, nylon 6 film,
and vinyl chloride-vinyl acetate copolymer film are approved for
use in radiation processing of prepackaged foods at a dose not to
exceed 60 kGy of gamma or X-radiation. The regulation (1) must be
consulted for details on the materials approved. Neither laminated packaging materials nor the use of electron irradiation are
included in the regulation.

Radiation Effects on Individual Packaging Materials

     General Radiation Chemistry of Polymers. Ionizing radiation
has two major effects on polymers: crosslinking and scission of the
polymer chains (2). Both reactions may occur simultaneously and
the predominating reaction dictates whether the polymer is degraded
or if it increases in molecular weight as well as undergoing changes
in its physical properties (2-5). The scission reactions may also
result in the production of hydrogen gas, hydrocarbons, carboxylic
acids, and changes in extractives from the various polymers.
Ionizing radiation can be effectively used to modify the properties
of polymers and improve their packaging properties (6-8). The
properties of ionizing radiation may, however, generate long-lived
free radicals (9) in the packaging materials which could conceivably
contribute to subsequent reactions in the packaging material or
presumably even in the food. The tests that must be applied to a
food packaging product include some that would not be applied to
polymers intended for use in medical applications. The physical
tests that must be applied however are similar and several common

tests applied to polymers intended for medical applications were described by Landfield (10).

Acrylic Plastics

The acrylic resins are polymers of acrylic or methacrylic esters and are readily polymerized, either as homopolymers or copolymers, with many other monomers (11). Polyacrylic esters, polyacrylic acid, polyacrylamide, butadiene-acrylonitrile copolymers, and styrene-acrylonitrile copolymers were listed by Bovey (2) as being among the polymers in which crosslinking rather than degradation would be the predominant reaction. Polymethyl methacrylate, polymethacrylic acid, and polymethacrylamide, on the other hand, were listed as among those polymers in which the predominant reaction with ionizing radiation would be degradation (2). Landfield (10) indicated that acrylonitrile-butadiene-styrene was stable for a single dose of 25 kGy. The regulation (1) includes coatings for polyolefin film or polyethylene terephthalate comprising a vinylidene chloride copolymer containing a minimum of 85% vinylidene chloride with acrylic acid, acrylonitrile, methyl acrylate, and methyl acrylate for radiation doses not to exceed 10 kGy. The flexible multilayer packaging used by the US Army for studies of irradiation sterilized beef and poultry products contained an ethylene-acrylic acid copolymer between the inside polyethylene layer and the aluminum foil central layer (12).

Cellulose and Cellulose Derived Products

Wax-coated paperboard is approved, as mentioned above, for use up to a radiation dose of 10 kGy (1). Ionizing radiation results in reduction of the crystallinity and depolymerization of the cellulose chain (13). Wierbicki and Killoran (14) reported that essentially no water, acetic acid, or n-heptane solubles were produced from vegetable parchment (KVP Satherland Paper Co.) after ionizing radiation doses of 60 kGy. Killoran (13) reported over two-fold reductions in the intrinsic viscosity of fiberboard and bleached sulfite paperboard after a radiation dose of 30 kGy. The tensile strength and tear resistance of the bleached sulfite paperboard was reduced by 19% and 7% in the machine and cross directions, respectively (13). Both electron and gamma irradiation of fiberboard resulted in decreased bursting strength. The changes observed in these cellulosic products upon irradiation were to be expected in that a large body of evidence exists documenting the degradation of cellulose by ionizing radiation (2). Bovey (2) cites studies from as early as 1929 that paper becomes brittle and crumbly and gives off hydrogen, carbon dioxide, and carbon monoxide when exposed to cathode rays. Glegg and Kertesz (15) describe the results of 60 kr to 2,300 kr doses on cellulose in which there was both an immediate and a delayed decrease in the intrinsic viscosity of the cellulose. Predominate chain scission occurs in the substituted as well as in the unsubstituted cellulose chain (2).

Vegetable parchment and nitrocellulose coated cellophane were
included in the same petition to the food and Drug Administration
from the Atomic Energy Commission (16). Zehnder (17) concluded
that derivatives of cellulose, such as cellophane, are not suitable
for packaging goods that are to be subjected to irradiation treatments.

Epoxy Plastics

These groups of thermo- or nonthermo- plastics contain epoxy
groups and are usually as coatings or adhesives (18). The laminated packaging material used by the U.S. Army for packaging of
electron sterilized beef and poultry products contained an epoxy
modified polyester as the adhesive between the outside nylon 6
layer and middle aluminum foil layer (12).

Ethylene-Vinyl Acetate

Patients exist for crosslinking of ethylene-vinyl acetate copolymers (19-21), and Chapiro (22) discusses in detail the radiation
induced polymerization of vinyl acetate and a radiation induced
copolymer graft of vinyl acetate-cellulose diacetate.

Glass

Most glasses are discolored when exposed to ionizing radiation
and turn brown (23).

Ionomers

No information was located concerning the reaction or lack of
reaction of Surlyn A or any other ionomer with ionizing radiation.

Nylon

Nylon is the generic name for long-chain polyamides and as mentioned above, nylon 6 and nylon 11 are included in the list of
films approved for irradiation of prepackaged foods (1). Eighteen
commercially available flexible films were evaluated for the
packaging of radiopasteurized fishery products by Tinker et al.
(25). The criteria used to evaluate the films were the organoleptic
qualities of the products and the seal efficiency and resistance to
bacterial penetration of the films. Using these criteria, polyethylene, polypropylene, saran coated cellophane, cellophane, and
nylon 6 were considered unsatisfactory. Nylon-11, saran-coated
nylon-11, polyolefin-coated polyester, polyethylene-coated polyester, laminated paper-aluminum-polyelefin-coated polyester,
laminated aluminum-paper-polyolefin coated-polyester, laminated
saran-polyethylene nylon, and saran-coated polystyrene lids and
trays were considered to be excellent for packaging radiopas-

teurized (2.5 kGy, sources ^{60}Co, temperature and dose rate not
stated) fish for a refrigerated shelf life of one month.
Bovey (2) reviewed the radiation chemistry of nylon up to 1958.
The evidence indicated that crosslinking was the predominate
reaction and that the elastic modulus and tensile strength
increased while elongation and impact strength decreased (2).
Keay in 1968 (26) concluded that nylon 11 was superior to two
laminated packaging materials for the packaging of irradiated
fish. Essentially no changes were found in the nylon 11 up to
doses of 160 kGy (26).

Polycarbonates

Two studies report results obtained with laminates in which polycarbonate was the food contacting layer (13, 14). Though neither
study discussed the polycarbonate film in any detail, the results
would seem to indicate that there were no marked changes in extractives (14) nor in tensile, burst, or seal strength (13) of the
laminates.

Polyesters

Bovey (2) reviews several studies of the effects of ionizing
radiation on polyesters stating that polyethylene terephthalate
has received the greatest attention. Bovey (2) did not feel that
the data conclusively indicated that scission predominated.
Chapiro (22) concluded in his review that data existed conclusively
demonstrating that polyethylene terephthalate was crosslinked by
ionizing irradiation. Killoran (13) investigated the properties of
polyethylene terephthalate following either electron or gamma
radiation doses of approximately 64 kGy. No radiation-induced
extractives were found in water, acetic acid, or n-heptane. No
significant effect was detected of irradiation on the tensile
strength, burst strength, or tear resistance of this polyester.
Cooper and Salunkhe (27) concluded that mylar bags were superior
to polyethylene bags for packaging of irradiated bing cherries.

Polyethylene: General Radiation Chemistry

The radiation chemistry of polyethylene has been extensively
investigated (2) and the predominate reaction is usually crosslinking. The petitions for the use of polyethylene were filed in
1964 by the Atomic Energy Commission for doses up to 10 kGy (16)
and by the U.S. Army for doses up to 60 kGy in 1967 (24). Radiation-induced crosslinking can be used to extensively modify the
properties of polyethylene for shrink wrapping (7) or for greatly
increased heat stability (2). Polyethylene reacts to radiation by
at first becoming increasingly insoluble due to the crosslinking
and with greater doses changes in color finally to ruby red (28).
Physical properties including flexibility and oxygen transmissibility may be altered (28).

Polyethylene Film: Extractives

Killoran (13) investigated the extractives from medium density
polyethylene irradiated in the presence of food-simulating
solvents. When either distilled water or acetic acid was used
as the food-simulating solvents, no identifiable changes were produced in total extractives or chloroform-soluble extractives by a
radiation dose of approximately 60 kGy from either gamma or
electron radiation sources. The gamma radiation temperature was
23-65°C, and the 10 MeV electron radiation temperature was 25-40°C.
The dose rates were 8 Gy/sec and 2×10^7 Gy/sec for gamma (^{60}Co)
and electron beam, respectively. Minimal change occurred in the
extractive from gamma irradiated medium density polyethylene in the
presence of n-heptane from that of the control. Less n-heptane
extractive was obtained from electron irradiated polyethylene than
from the control. The irradiation produced no chloroform-soluble
residues from the polyethylene in water or acetic acid, and the
n-heptane soluble residue was identical to that from the control.

Polyethylene Film: Radiation Induced Volatiles

Killoran (13) reported that electron irradiation of low-density and
high-density polyethylene at average radiation doses of 10, 60,
and 120 kGy produced hydrogen, methane, hydrocarbons, and carbon
dioxide. The quantities of volatiles increased with increasing
radiation dose; almost 3 times the total amount of hydrocarbons
were produced by a radiation dose of 60 kGy as opposed to a dose
of 10 kGy from the low-density polyethylene. Some ninety hydrocarbons were identified which ranged in molecular weight from
$16(CH_4)$ to $184(C_{13}H_{28})$. Almost twice the amount of hydrocarbons
were produced from the low as opposed to the high-density polyethylene.
          In a much more recent series of studies, Azuma et al. (29-30)
identified the odor-producing volatiles from electron beam and
gamma irradiation sterilization of low density polyethylene films.
Azuma et al. (29), using contemporary gas chromatographic and gas
chromatographic-mass spectroscopic techniques, identified the
volatile compounds produced from six types of low-density polyethylene films by a dose of 20 kGy from a 2.5 MV electron beam.
Aliphatic hydrocarbons accounted for approximately 35% of the total
peak area with saturated hydrocarbons up to C_{13} predominating.
Four aldehydes (C_2-C_5) and six ketones (C_4-C_8) accounted for
approximately 26% of the total peak area. Five carboxylic acids
(C_2-C_5) accounted for 18% of the peak area, and small amounts of
alcohols, toluene, and phenol were identified. The odor from a
mixture of the identified components in the same volume ratio as
their peak areas resembled that of the off-odor of the irradiated
polyethylene. Azuma et al. (30) studied the effects of film
variety on the amounts of carboxylic acids produced by electron
irradiation of polyethylene film. Nine types of low-density
polyethylene were examined after a radiation dose of 20 kGy; and
acetic, propionic, n-butyric, and n-valeric acids were quantitated.
The predominate product (40 to 75% of total carboxylic acid) was

acetic acid followed by propionic acid (30). The addition of
butylated hydroxytoluene to the film reduced the total amounts of
carboxylic acids that were formed. It should be noted that the
total amounts of the four carboxylic acids produced by the irradiation of the films ranged from 1.83 μg/g to 15.7 μg/g.
Azuma et al. (31) extended their studies to include the effects of
the conditions of electron beam irradiation and a comparison of
electron-beam to gamma-irradiation on the production of volatiles
for irradiated polyethylene film. At an irradiation dose of 20 kGy
(2.5 MeV e⁻, beam current 250 μA), Azuma et al. (31) noted that the
formation of carbonyl compounds increased several fold as the
oxygen content was increased from zero to five percent. Hydrocarbon formation was not noticeably affected by the oxygen content.
The production of both carbonyls and hydrocarbons during electron
irradiation of the polyethylene film could be lowered dramatically
by irradiation at temperatures below polyethylene's glass transition temperature of about -78°C. The amount of carboxylic acids
that were formed by a radiation dose of 20 kGy from 2.5 MeV e⁻ at
a beam current of 250 μA were relatively constant at temperatures
above 0°C but decreased to 35% at -75°C and to 16% at -196°C.
Azuma et al. (31) discovered that there was a sharp increase in
the production of carboxylic acids from polyethylene when the
energy was increased from 1.5 to 2.0 MeV. Unfortunately from the
viewpoint of this reviewer, the investigators did not extend their
study up to a level of 10 MeV, which would be required for the
irradiation of many foods. They also did not state the other
irradiation conditions, e.g., temperature, or atmosphere. It was
noted that lower beam currents, i.e., 125 μA, increased the formation of carboxylic acids (31). The formation of carboxylic
acids almost doubled when the polyethylene received a dose of 20
kGy from ^{60}Co rather than from the 2.5 MeV electron beam source.
Unfortunately, the authors did not indicate the dose rate or
temperature of the gamma irradiation treatment. Azuma et al. (31)
interpreted these results as being due to dose rate. With gamma
radiation, the probability of crosslinking would be low because of
a low concentration of primary radicals and a high relative availability of oxygen. With electron beam radiation, the recombina-
tion of the primary radicals, rather than their oxidation would be
more favored because of their much higher relative concentration
(31).

Polyethylene-Radiation Induced Crosslinking

The effects of ionizing radiation on crosslinking of polyethylenes
and grafting of polyethylene and polypropylene to other materials
has been the subject of many studies (6). Generally these studies
appear to have limited application to our consideration of the
effects of ionizing radiation on materials used for packaging of
foods; but nevertheless, these studies do provide a strong theoretical background and may, as well, provide the basis for
radiation-stable materials suitable for packaging foods that are to
be subjected to irradiation treatments. This field has moved with
such rapidity that Baird and Joonase (32) listed a bibliography of

500 patents related to radiation crosslinking of polymers in 1982.
Godlewska et al. (33) described the use of ionizing irradiation to
increase dramatically the impact resistance of polyethylene films
intended for use in packaging foods. Azuma et al. (30) described
the effects of incorporating an antioxidant into low-density
polyethylene on the subsequent generation of carboxylic acids from
the film during irradiation. Gal et al. (34) described the effects
of three antioxidants on the radiation crosslinking efficiency of
low-density polyethylene. All three antioxidants that were investigated decreased the amount of crosslinking at a given dose.
Schlein et al. (35) obtained a patent for the addition of
2,2'-methylene-bis(4-ethyl-6-t-butyl) alcohol to polyethylene,
which was reported to stabilize it so that there was no odor formation at sterilization doses for meats (20-40 kGy).

Polypropylene

Bovey considered the predominant reaction of ionizing radiation
with polypropylene to be that of crosslinking (2). Chapiro (22)
discussed the nature of the crosslinking reactions. Tinker et al.
(25) considered polypropylene to be unsuitable for the packaging of
radiopasteurized fish because the gas permeability rate was too
high. Varsanyi and Farkas (36) concluded that non-oriented
polypropylene film was suitable at dose levels of less then
8.0 kGy, did not suffer significant alteration of either chemical
or physical properties, and was suitable for packaging of radiopasteurized meats. The radiation-crosslinking of polypropylene
to improve its heat resistance and tensile strength was described
by Benderly and Bernstein (37).

Polystyrene

Bovey reviewed several studies on the effects of ionizing radiation on polyestyrene concluding that the predominant reaction was
crosslinking (2). Very little gas was formed during the crosslinking. Kline (38) reported the results of studies of radiation
on the dynamic mechanical properties of styrene concluding that
crosslinking took place but no changes in the dynamic mechanical
properties were detected. Chapiro (22) provided an extensive
review of the effects of ionizing radiation on polystyrene.
Tinker et al. (25) reported polystyrene lids and trays to be good
to excellent for the packaging of radiopasteurized fish.

Polytetrafluoroethylene

Polytetrafluoroethylene, though almost chemically inert, is degraded
by ionizing radiation. It loses mechanical strength and CF_4 is
evolved (2).

Polyvinyl Chloride

Bovey (2), in reviewing the existing studies as of 1958, indicated
that vinyl chloride polymers were on the border line between

predominant formation of crosslinks and/or scission reactions.
Several studies were cited as indicative of degradation, including
the evolution of hydrogen chloride, darkening, reduction in tensile
strength, and hardness. Some reports, however, cited by Bovey (2)
indicated that crosslinking was taking place under some circum-
stances. Chapiro (22) reported in 1962 that his research indicated
that irradiation in air caused a steady degradation of the polymer,
but that irradiation in vacuo led to only minor changes at doses
as high as 83 kGy. Yegorova et al. (39) reported that radiation
crosslinking predominated in the case of the ethylene copolymer
containing 15 mole % vinyl chloride. The tensile strength was
reported to rise with increasing dose. Haesen et al. (40)
reported that gamma irradiation stimulated the formation of a
heptane soluble, nonvolatile Sn compound with a high migration
tendency from organo-tin compounds added to poly(vinyl chloride)
for stabilization properties.

Other Plastic Films

Killoran (13) tested eight plastic films for water, acetic acid,
and n-heptane extractives following either gamma or electron
radiation to an average dose of 63 or 67 kGy, respectively.
Polyethylene-polyisobutylene blend, plasticized polyvinyl chloride, polyethylene terephthalate and polystyrene had minimal net
changes in water or acetic acid extractives. Four other gamma-
irradiated films, polyiminocaproyl, polyiminoundecyl, poly(vinyl
chloride-vinyl acetate), and poly(vinylidene chloride-vinyl
chloride) had increased water and acetic acid extractives after
irradiation. There were minimal changes in n-heptane extractives
after gamma- and electron-irradiation of polyiminocaproyl and
polyethylene terephthalate. Increased amounts of n-heptane
extractives were obtained from gamma- or electron-irradiated films
of polyethylene polyisobutylene blend, plasticized polyvinylchloride, poly(vinylidene chloride-vinyl chloride) and polystyrene
than from non-irradiated controls. Gamma-irradiated films of
polyethylene, and polyiminoundecyl and electron-irradiated
poly(vinyl chloride-vinyl acetate) also had increased n-heptane
extractives. There were minimal changes in n-heptane extractives
after gamma-irradiation of poly(vinyl chloride-vinyl acetate), and
electron-irradiation of polyiminoundecyl. Electron-irradiated
medium-density polyethylene had less extractive than the non-
irradiated control. The author attributed the differences in
extractives to the relative stability of the films to crosslinking
and/or degradation by either the gamma radiation or electron
radiation. Killoran (13) concluded that the radiation stability,
based on the total amounts of volatiles produced by electron
irradiation of each film, was polyiminocaproyl > high-density
polyethylene > poly(vinylidene chloride-vinyl chloride) > low-
density polyethylene. The abrasion resistance of low-density
polyethylene and polyiminocaproyl increased with increasing
radiation dose (13).

Laminated-Packaging

Keay (26) examined the effects of gamma radiation up to a dose of
160 kGy from spent fuel elements (no temperature, atmosphere, or
dose rate was stated) on pouches of nylon 11, pouches of laminated
polypropylene (0.002 inch thick)-medium density polyethylene
(0.002 inch thick) bonded with an unspecified adhesive, and laminated pouches of 0.002 inch medium-density polyethylene and
(0.0005 inch) polypropylene bonded with an unspecified adhesive.
The evaluations included development of color or change in transparency, loss of slip, brittleness, delamination, infrared
spectral examination, gas chromatographic examination, and a taste
panel examination of cod fillets stored in the vacuum sealed
irradiated pouches at 2°C for 5 days. The fish fillets themselves
were not irradiated. They were evaluated for raw odor taint and
then steam cooked and tested for odor and flavor taint. The nylon
11 was essentially unaffected by the radiation treatments through
a dose range of 80 kGy. Both of the laminates, however, had dose-
related increases in odor though only the polypropylene-polyethylene
pouches had a loss of slip with increasing dose. Both laminates
were discolored by the radiation. The polypropylene-polyethylene
pouches had a four-fold increase in total volatiles after a radiation dose of 80 kGy and strongly tainted the raw but not the cooked
fish fillets (26).
        Killoran et al. (41) investigated the use of five commer-
cially available plastic laminates for the packaging of bacon, ham,
or pork, which was to be irradiated at 3°C to a dose of 45 to 56
kGy at a dose rate of 53 krads per min using a ^{60}Co radiation
source. The evaluation included observations for odor, leakage,
color changes, and determination of the physical changes in the
pouches. The following packaging materials were evaluated: Film
A: 0.5 mil 50 A Mylar/0.5 mil Al foil/2 mil poly(vinyl chloride);
Film B: 30 lb paper/0.35 mil Al foil/2 mil Scotchpak 20A5; Film C:
30 lb paper/0.35 mil Al foil/1 mil Marlex TR-515; Film D: 0.3 mil
Al foil/30 lb paper/2 mil Scotchpak 20A5; and Film E: 4.5 mil
transparent Scotchpak 45A27. The last named component was in each
case the food-contacting film. Only film B proved satisfactory
for packaging of the three meat products over a one-year period.

Packaging Studies of Shelf-stable Meat and Poultry

The most comprehensive published studies of the effects of ionizing
radiation on packaging materials were conducted at the U.S. Army
Laboratories, Natick, Massachusetts to develop packaging materials
suitable for use in packaging irradiated foods intended to have a
two year shelf life at room temperature. Large scale studies of
the wholesomeness of radiation treated (to commercial sterility)
poultry and beef products were conducted. The products used for
these studies received radiation doses in excess of 40 kGy at an
average temperature of -25 ± 15°C and the studies were conducted
on such a scale as to provide realistic testing of the packaging
materials. In the most recent study, 135,405 kg of broiler
chicken meat was processed and packed either in cans or in flexible

packages depending on the treatment it was to receive (42-44).
The details of the processing of the chicken were reported by
Wierbicki (44).

Tinplate Cans

The chicken that was to be sterilized by gamma radiation was vacuum
packed in 404 x 309 mm epoxy-phenolic- enamel lined cans (42-44).
The cans were constructed of 80 to 90 weight, No. 25 tinplate,
coated with epoxy-phenolic enamel (44). The lids were sealed with
a blend of cured and uncured isobutylene-isoprene copolymer
(43-45). Killoran et al. (45) had evaluated the tinplate can with
certain enamels and end-sealing compounds and found it to be
suitable for packaging meat and poultry products that were to be
irradiated to a dose of 70 kGy at -60°C.

Flexible Packaging.

The chicken that was to be sterilized by electron beam was packed
in flexible packages 165 mm x 208 mm and fabricated with 0.025 mm
polyiminocaproyl (nylon 6) as the outside layer, 0.0090 mm aluminum foil as the middle layer, and 0.062 mm polyethylene
terephthalate-medium density polyethylene as the food contacting
layer (42-44). The actual food contactant was the medium density
polyethylene layer. The reliability of this flexible packaging
material was extensively tested during the packaging of the beef
described above (12). Under conditions that would have been
similar to commercial use, only 140 of 441,470 pouches containing
beef were defective after the vacuum sealing operation for a defect
rate of 0.03%. The bond strength of the pouches increased
approximately 3 fold during the irradiation process and did not
deteriorate during storage for 24 months (12). The burst strength
of the pouch also increased during the irradiation process from
1.9 x 10^5 Pa to 2.4 x 10^5 Pa (12).
    The food-containing layer in the pouches used for packaging
the chicken was extensively tested for possible production of
extractives during irradiation by Killoran (43). Standard food
simulating solvents were used: water, acetic acid (pH 3.5), and
n-heptane. The film was formed into pouches that would hold 160 g
of food and had an exposed surface area of 290.3 cm^2 per pouch.
Five pouches were used for each sample. The pouches with food-
simulating solvent were sealed in an atmosphere of nitrogen,
cooled to -40°C, and irradiated with 10 MeV electrons at a dose
rate of 2 x 10^7 to 5 x 10^7 kGy/sec, to a total dose of 47 to
71 kGy. The initial temperature, as previously mentioned, was
-40°C and rose to -18°C during the irradiation. After thawing
the pouches were stored at 50°C. Under these conditions of
irradiation, the total water extractives were 0.54 mg/pouch and
0.33 mg/pouch when not irradiated. When the equilibrium values
for the extractives in pH 3.5 acetic acid were corrected for that
formed in the acetic acid when irradiated, (1.17 mg/pouch) the
irradiated samples had 0.26 mg/pouch and the non-irradiated

0.40 mg/pouch. In a similar manner, when the equilibrium extractives for n-heptane were corrected for radiation induced
extractives in the heptane (1.35 mg/pouch), 0.57 mg/pouch and
0.76 mg/pouch were found in the irradiated and the non-irradiated
samples, respectively. The grease-like material extracted from
the pouch by n-heptane was not part of the adhesive between layers,
which was a cured polyester-epoxy system. It was identified by UV,
IR, and mass spectra as a low molecular weight polyethylene.

Conclusion

Data exists on the reactions of ionizing radiation with a number
of polymer and copolymer films. Some of these films such as
polyethylene, polystyrene, and polyethylene terephthalate appear to
be very well suited for use as packaging materials for foods which
are to be subjected to ionizing radiation. But even these materials do react with ionizing radiation and it should be remembered
that the regulation (1) was written from data submitted in the
early 1960s. Modern analytical technology should be applied to the
analysis of volatiles and extractives. Few laminated films have
been extensively investigated and none have been approved for use
in packaging foods which are to be irradiated. The use of ionizing
radiation to improve the characteristics of films opens the
question as to what subsequent additional reactions might take
place if such films were used to package foods which were to be
irradiated. Fortunately, the extensive use of radiation crosslinking by industry should provide the knowledge to select the
proper films for formation of laminated packaging materials to
provide safe and reliable materials for use in packaging foods
which are to be irradiated. It is possible that satisfactory
packaging films could be produced by co-extrusion of polymers which
have been previously approved avoiding the necessity of seeking a
new approval from the Food and Drug Administration. The approvals
already granted for irradiation of foods and the potential for
additional approvals should generate a considerable market for
packaging materials cleared for use with ionizing radiation.
Abrreviations used: Gray(Gy) - A radiation dose of 1 Gy involves
the absorption of 1 J of energy per kilogram of matter.
1 Gy = 100 rad. 1 kGy = 1000 rad. Rep(r) - Roentgen equivalent
physical. An absorbed dose of ionizing radiation equivalent to 93
ergs/g; 1 Gy = 107.5 rep, 1 million electron volts (MeV).

Literature Cited

1. Title 21--Food and Drugs Subpart C--Packaging Materials
    for Irradiated Foods 179.45, Code of Federal Regulations
    (U.S.), 1986, p 356.
2. Bovey, F. A., The Effects of Ionizing Radiation on Natural and
    Synthetic High Polymers; Interscience Publishers, Inc.:
    New York, NY, 1958.
3. Charlesby, A., Nature 1953, 171, 167.
4. Lawton, E. J., Bueche, A. M., Balwit, J. S., Nature 1953,
    172, 76-77.

5. Miller, A. A., Lawton, E. J., Balwit, J. S., J. Polymer Sci. 1954, 14, 503-504.
6. Chapiro, A., Radiat. Phys. Chem. 1983, 22, 7-10.
7. Benning, C. J., Plastic Films for Packaging; Technomic Publishing Co., Inc., Lancaster, PA, 1983.
8. Roediger, A. H. A., Du Plessis, T. A., Radiat. Phys. Chem. 1986, 27(6), 461-468.
9. Hollain, G. de, J. Indust. Irradiation Tech. 1983, 1(1), 89-103.
10. Landfield, H., Radiat. Phys. Chem. 1980, 15, 39-45.
11. Whittington, L. R., Whittington's Dictionary of Plastics; Technomic Publishing Co., Inc., West Port, CT, 1978.
12. Killoran, J. J., Cohen, J. J., Wierbicki, E., J. Food Processing and Preservation 1979, 3, 25-34.
13. Killoran, J. J., Radiation Research Reviews 1972, 3, 369-388.
14. Wierbicki, E., Killoran, J. J., Research and Development Associates, Activities Rept. 1966, 18(1), 18-29.
15. Glegg, R. E., Kertesz, Z. I., J. Polymer Sci. 1957, 26, 289-297.
16. Federal Register 1964, 29, 11651.
17. Zehnder, H. J., Alimenta 1984, 23(2), 47-50.
18. Wheaton, F. W., Lawson, T. B., Processing Aquatic Food Products; John Wiley & Sons, New York, 1983, Chapter 12.
19. Oyama, M., Ohkubo, S., Sakai, T., Osaka, D., Jap. Patent 70 15 137, 1970, Chem. Abstr. 1973, 73, 99651e.
20. Tubbs, R. K.   U.S. Patent 3 734 843, 1973.
21. Brax, H. J., Porinchak, J. F., Weinberg, A. S.   U.S. Patent 4 278 738, 1981.
22. Chapiro, A., Radiation Chemistry of Polymeric Systems; Interscience Publishers, New York, 1962.
23. Elias, P. S., Chemistry and Industry 1979, 10, 336-341.
24. Federal Register 1967, 32, 8360.
25. Tinker, B. L., Ronsivalli, L. J., Slavin, J. W., Food Technol. 1966, 20(10), 122-124.
26. Keay, J. N., J. Fd. Technol. 1968, 3, 123-129.
27. Cooper, G. M., Salunkhe, D. K., Food Technol. 1963, 17(6), 123-126.
28. Briston, J. H., Katan, L. L., Plastics Films; Longman Inc., New York, 1983 2nd ed, p 16.
29. Azuma, K., Hirata, T., Tsunoda, H., Ishitani, T., Tanaka, Y., Agric. Biol. Chem. 1983, 47, 855-860.
30. Azuma, K., Tanaka, Y., Tsunoda, H., Hirata, T., Ishitani, T., Agric. Biol. Chem. 1984, 48, 2003-2008.
31. Azuma, K., Tsunoda, H., Hirata, T., Ishitani, T., Tanaka, Y., Agric. Biol. Chem. 1984, 48, 2009-2015.
32. Baird, W., Joonase, P., Radiat. Phys. Chem. 1982, 19, 339-360.
33. Godlewska, E., Kubera, H., Skrzywan, B., Jaworska, E., Opakowanie 1984, 30(4), 25-29.
34. Gal, O. S. Markovic, V. M., Novakovic, L. R., Radiat. Phys. Chem. 1985, 26, 325-330.
35. Schlein, H. N., La Liberte, B. R.   U.S. Patent 3 194 668, 1965.

36. Varsanyi, I., Kiss, I., Farkas, J., Acta Alimentaria 1972, 1, 5-16.
37. Kline, D. E., J. Appl. Polymer Sci. 1969, 13, 505-517.
38. Kline, D. E., J. Appl. Polymer Sci. 1961, 5(14), 191-194.
39. Yegorova, Z. S., Yel'kina, A. I., Duntov, F. I., Karpov, V. L., Leshchenko, S. S., Vysokomol. soyed 1969, A11(8), 1766-1773.
40. Haesen, G., Depaus, R., Tilbeurgh, H. van, Le Goff, B., Lox, F., J. Indust. Irradiation Tech. 1983, 1(13), 259-280.
41. Killoran, J. J., Breyer, J. D., Wierbicki, E., Fd. Technol. 1967, 21(8), 73-77.
42. Thayer, D. W., Christopher, J. P., Campbell, L. A., Ronning, D. C., Dahlgren, R. R., Thomson, G. M., Wierbicki, E., J. Food Protection 1987, 50, 278-288.
43. Killoran, J. J. Packaging materials for use during the ionizing irradiation sterilization of prepackaged chicken products; ERRC-ARS Document No. 82. PB84-186998 1984. National Technical Information Service, Springfield, VA.
44. Wierbicki, E. Food Irradiation Processing, Proc. Symp. Washington, DC. March 4-8, 1985. International Atomic Energy Agency. Vienna, Austria.
45. Killoran, J. J., Howker, J. J., Wierbicki, E. J., Food Processing and Preservation 1979, 3, 11-24.
46. Killoran, J. J., Wierbicki, E., Pratt, G. P., Rentmeester, K. R, Hitchler, E. W., Fourier, W. A. In Chemistry of Food Packaging; Advances in Chemistry Series No. 135; American Chemical Society: Washington, DC, 1974, p 22-34.

RECEIVED September 24, 1987

Chapter 16

Relationship Between Polymer Structure and Performance in Food Packaging Applications

George W. Halek

Food Science Department, Rutgers University,
New Brunswick, NJ 08903

This paper reviews the performance of polymers as food packaging materials from the viewpoint of the chemical and physical structure of the polymers and chemical and physical interactions with food ingredients. The major factors involved in these interactions are polarity and crystalline structure of the polymer and polarity and chemical structure of the food ingredients. The effects of these factors on mechanical, barrier, and compatibility behavior of the systems are described, and the principles controlling the behaviors are applied to explain performance. Recommendations are given for further research to broaden understanding of polymer-food interactions.

The performance that is to be expected from a polymeric food packaging material is the same as that expected from any food packaging material. It has been described elswhere (1,2) in detail and can be summarized for our purpose as: containing the food, protecting it from the environment, and maintaining food quality. In the case of polymers, the ability to perform these functions will depend on their mechanical and barrier properties, to which must be added the requirement of long-term compatibility with the food. This last property will depend on the nature of the food ingredients and the structure of the polymeric packaging material. Thus food packaging performance is seen to depend on polymer mechanical, barrier, and compatibility properties. These, in turn, will depend on polymer structure and changes that can occur with time during interactions with the food ingredients.

As a practical matter, there are now a number of polymers that are used in food packaging applications (3), and they have endured because they have demonstrated a defineable level of appropriate mechanical and barrier properties for the task. The matter of compatibility has been less easy to define and in an increasing number of cases has been recognized as a potential source of loss in food quality. A study of the literature has revealed only a limited amount of publication of such cases, but there are enough to permit categorization. These will be described in this paper.

0097-6156/88/0365-0195$06.00/0
© 1988 American Chemical Society

POLYMER STRUCTURE CONSIDERATIONS. The polymers that are used in food packaging applications can be divided into several classes according to their chemical structure (Table I).

Table I. Classes of Food Packaging Polymers

Classes	Representative Polymers
Hydrocarbon	Polyethylene, Polypropylene, Polystyrene.
Halogen	Polyvinyl chloride, Polyvinylidene chloride.
Functional Vinyl	Polyvinyl alcohol, Polyvinyl acetate, Polyacrylonitrile.
Condensation	Nylon 6,6, Polyethylene terephthalate.
Miscellaneous	Cellophane, Polycarbonate, Polyurethane.

Each of these polymers can next be assigned to a series of increasing polarity ranging for example from polyethylene (low polarity) to polyvinyl alcohol (high polarity) due mainly to the differences in the functional groups attached to the main polymer chain. They can also be arranged according to crystalline/amorphous solid state ratios. For example they range from isotactic polypropylene with a high crystalline/amorphous ratio to polyvinyl choride with a low ratio. Both of these factors, degree of polarity and crystalline/amorphous ratios, are major factors in the state of matter attained in a polymeric packaging material. That attained state of matter controls the properties of the fabricated packaging materials, and this depends on polymer structure (4). In addition there are a number of other factors that affect performance such as molecular weight and its distribution, crosslinking, and additives, but these are secondary to the polarity and crystallinity factors dealt with here. However these effects should not be ignored in seeking information on causes of polymer performance.

FOOD PACKAGING PERFORMANCE CONSIDERATIONS. The major performance characteristics of polymeric food packaging materials controlled by the structural considerations are shown in Table II. The table also includes typical properties that are measured for those performance characteristics.

Table II. Performance Characteristics and Related Properties

Performance Characteristic	Related Properties
Mechanical Strength	Tensile Strength
	Compressive Strength
	Impact Strength
Barrier Behavior	Permeation
	Migration
Compatibility	Sorption
	Desorption

These performance characteristics and how they are related to polymer structure and the effects on food packaging applications are presented in the next sections.

POLYMER STRUCTURE AND MECHANICAL PERFORMANCE. The variety of polymer classes that are used in food packaging is shown in Table I. They represent the range of functional groups and crystalline structures that provide properties that meet the needs of food packaging applications. The polymers of low polarity for example the polyhydrocarbons, attain their mechanical strength from packing of very long chains of high molecular weight or regularity so that the relatively weak van der Waals interchain forces add up sufficiently to meet the material requirements. Those of higher polarity can attain that total requirement of interchain attractive force at a lower molecular weight and with less structural regularity because of the compensating polar attractive forces. These factors are well known in polymer work (4) and will not be described further. We will focus on the forces involved in food-polymer interactions that can change mechanical properties. This involves transitions in the polymer caused by migration of ingredients into the polymer in contact with food.

It has been recognized for some time that there is a strong dependence between diffusion and relaxation in polymers (5) and that this could be explained by a coupled diffusion-relaxation process (6,7). The argument is that when an organic penetrant is sorbed by a semicrystalline polymer, the rate of sorption is controlled by diffusion and a slow relaxation of the polymer (5). This further enhances the rate of diffusion which further influences the relaxation. To the extent that this occurs, changes occur in the polymer's properties. There are few published works covering the details of such behavior resulting from food-polymer interactions, but the effects expected can be summarized from general principles involving coupled diffusion-relaxation. In general, mechanical properties will be changed by permeants that can affect polymer chain segmental mobility. Accordingly, food ingredients that are absorbed and plasticize the polymer chains will permit them to slip past each other. Food ingredients that can displace polar attractive forces (hydrogen bonds or polar bonds) in the polymer will also affect chain mobility. The net effects of these two factors would be a lowering of tensile and compressive strength and modulus, and an increase in creep properties resulting from continued deformation with time. Impact strength will usually be increased under these conditions as the increased segmental motion and decreasing crystallinity would permit dissipation of impact energy by energy-consuming contortions of the more mobile polymer chain segments. Thus, the mechanical properties are affected significantly by the behavior of the polymer towards permeation by the food ingredients. The structural features that affect this behavior will be further considered in the next section on barrier performance.

POLYMER STRUCTURE AND BARRIER PERFORMANCE. Summaries of the relationships between polymer structure and barrier performance in food packaging materials are available (2,8,9). Much of the emphasis has been on permeation of H_2O, O_2, and CO_2 which are of particular importance to food shelf life. In addition, there is a limited amount of data published on permeation by model solvents and aroma compounds. Our discussion will include principles that cover all of these cases.

In general, there are several polymer structural features that result in good barrier performance: 1) Some polarity but of certain types; 2) Regularity of molecular structure; 3) Close chain-to-chain packing in the solid state. The polarity is important from the viewpoint of functional group interactions with the polar permeants of interest to foods. The regularity of molecular structure affects the ability of the polymer chains to approach each other during the crystallization process of the packaging material fabrication steps. That factor and the final close chain-to-chain packing that it permits provides a restricted pathway for the permeants. The problem is that at times these factors can have cross-effects that must be taken into consideration when deducing the effects of structure on performance. Such effects in the case of polarity can be seen from the data in Table III derived from the literature (2,3,8,9) for O_2, CO_2 and H_2O permeation in a number of barrier polymers.

Table III. Barrier Properties of Selected Polymers (2,3,8,9)

Polymer[a]	O_2dry[b]	O_2wet[c]	CO_2[b]	H_2O[d]	Functional Class
PVOH	0.02	7.00	0.06	10.00	Hydroxyl
EVOH	0.05	7.00	0.23	10.00	Hydroxyl
PVDC	0.08	0.08	0.30	0.05	Halogen
PAN	0.03	0.03	0.12	0.50	Nitrile
PET	5.00	5.00	20.00	1.30	Ester
NYLON 6,6	3.00	15.00	5.00	24.00	Amide
PP	110.00	110.00	240.00	0.30	Hydrocarbon

a PVOH, polyvinyl alcohol; EVOH, ethylene vinyl alcohol copolymer; PVDC, polyvinylidene chloride; PAN, polyacrylonitrile; PET, polyethylene-terphthalate; Nylon 66, hexamethylenediamine adipic acid; PP, polypropylene.
b cc-mil/100in^2 . day . 1 atm
c cc-mil/100in^2 . day . 1 atm at 80% RH
d g-mil/100in^2 . day . at 100°F and 90% RH

These data illustrate the cross effects when humidity is included. For example, polyvinyl alcohol is an outstanding barrier towards O_2 and CO_2 when dry. The high content of OH groups permits interactions that retard transport through the polymer. However, when water enters the polymer it preferentially interacts with the OH groups and thus diminishes retardation of the permeants and also swells the structure to increase interchain distances and thus free volume. The net result is a strong reduction in barrier performance. The same behavior is noted for ethylene vinyl alcohol copolymer with its OH groups. A similar behavior is seen for Nylon 6,6 with its amide groups, so it is not restricted just to OH groups. The effect is not observed for PVDC or PAN or PET (see Table III) which also have polar groups. The indication is that if a permeant can preferentially interact with the polymer and can swell it, the barrier performance is reduced (PVOH, EVOH, Nylon 6,6). On the other hand, if it cannot, the barrier performance remains (PVDC, PAN, PET, PP). This is also noted for the water permeation data in Table III for these polymers. Those data show that the two polymers with OH groups and the Nylon are poor water barriers while the others are better. But, so is the performance of the non-polar polypropylene. Thus polar goups per se do not mean good or bad barrier performance towards water nor towards O_2 and CO_2. So long as the polymer polarity results in strong interchain attractions that are not displaced by the permeant, the barrier performance remains good.

The other factor of importance to barrier performance is attained crystallinity. The whole topic is complex but can be illustrated by the following examples from Table III and Table IV derived from the literature (2,3,8,9).

Table IV. Crystallinity Effects from Stretch Orientation

Polymer	O_2 Permeation[a]		H_2O Permeation[b]	
	Unoriented	Oriented	Unoriented	Oriented
PP	414.00	240.00	0.5	0.3
Nylon	2.60	1.30	25.0	8.0
PET	8.00	4.00	3.0	1.3
PAN	0.03	0.03		

a. cc–mil/100 in^2 . day . atm
b. g–mil/100 in^2 . day 90% RH at 38°C

The polymers in Table III are all highly crystallizable as a result of their conformational regularity which permits packing into specific crystal lattices during the melting and cooling process for film or container fabrication. Such crystallinity reduces the volume of amorphous polymer available as a permeation medium and also increases the pathlength to be travelled through the plastic structure. The result is an improvement in barrier properties from attained crystallinity in addition to the previously mentioned retardation effect resulting from functional group interaction with the permeant. The crystallinity effect can be further increased by stretch orientation of the finished structures. This is illustrated in Table IV for several polymers that can be stretch oriented. Orientation has improved the O_2 and H_2O barrier as a result of stretch treatment that has reduced the amount of amorphous phase and also diminished the distance between crystallites. The generalization cannot be made more specific here because of the numerous differences in organizational arrangements that are described in the literature (10, 11, 12). For example, in some cases the movement of the crystallites actually creates voids that diminish barrier effectiveness.

One further point of interest is that of laminates. Current technology permits formation of composites either by adhesive coatings or coextrusions to combine a desired set of properties in one structure. For example the outstanding O_2 and CO_2 barrier of PVOH can be combined with polypropylene which is four magnitudes poorer in that respect but which can provide the water barrier that the PVOH lacks. Guidance in interpreting the results to be expected from such combinations can be obtained by employing the basic principles described above.

POLYMER STRUCTURE AND COMPATIBILITY. In this section we are taking compatibility to mean the ability of the polymer and the food to coexist without significant changes occurring in either one as a result of their contact. There is a close relationship between this subject and the discussions in the section on barriers because the factors for interactions in the polymer are the same: polarity and crystallinity. However, here we will put more emphasis on the nature of the migrants. Further, we will limit our discussion to migration of ingredients from the food into the polymer. Migration in the other direction has been covered extensively by the large volume of publications involving regulatory concerns connected with migration of toxic ingredients or adulterants into foods (13,14). Even within those studies a body of theoretical considerations of migration behavior at low concentrations of migrants has been formed and the reader is directed to two articles (15,16) that discuss the main viewpoints involved. The whole area is quite extensive and will not fit into this presentation.

There are two main ways in which food can interact with polymers namely, 1) The food can react with or complex or in some way form a bond with the packaging material surface; 2) The food can be absorbed into the packaging material. In both cases the interactions depend on the nature of the food ingredients and on the polarity and state of matter of the polymer. A current search of literature on this subject reveals only a limited amount of publication involving measurement of food–polymer interactions. A sampling of recent publications is listed in Table V.

These examples, while restricted to polyolefins, indicate the kind of behavior observed for the sorption of food ingredients by polymers. Reference 17 showed that limonene was strongly sorbed by polyethylene while other oxygenated aroma compounds were sorbed to a lesser extent. The other papers also report equilibrium partitioning that is quite favorable to the plastic. For example the partition coefficient reported for limonene between polyethylene and water was 3,400 (19). For the other ingredients from that work, the coefficients ranged from 2–38. In another work (18) the sorption isotherms for limonene in water into polyethylene showed a biphasic sorption behavior.

Reaction rates indicated that adsorption was occurring very rapidly while absorption was slower. Carvone behaved similarly but had a much lower partition coefficient (18). Similarly, differences were found for the sorption rates of terpenes, linalool, and citronella with polyethylene and polypropylene (20). These reports confirm that the hydrocarbon model permeants are sorbed to a greater extent by polyolefins than are the more polar ingredients.

Table V.   Food Ingredient Sorption by Polymers

Ingredients	Polymers	Reference
Limonene	Polyethylene	17
Limonene	Polyethylene	18
Carvone		
Limonene	Polyethylene	19
Citral		
Benzaldehyde		
Ethyl Butyrate		
Terpenes	Polyethylene terephthalate/	20
Linalool	Foil/Polyethylene or poly-	
Citronella	propylene	
Fats in the form of	Polyethylene	14
ten foods including		
cheeses, margarine,		
chocolate, cream, milk		
and yogurt		

The data reported on fat sorption from dairy products (14) were
obtained by placing the foods, including solid structures, in
direct contact with the polyethylene film. The results showed a
wide range of sorption of the fats by the plastic that was
dependent on the structure of the foodstuffs. From these reports
there is not enough information to permit comment on the effects of
polarity or crystallinity on the polymer-food ingredient
interactions. Nor do they allow comment on differentiating surface
adsorption from internal absorption. However, they do show that the
general principles of sorption behavior established with non-food
ingredients are being followed.

FUTURE RESEARCH NEEDS.   From our brief review it can be seen that
there is a need for broadening the scope of research on the
interactions between food and packaging polymers. The broadening is
needed both in the nature of the food ingredients and the structure
of the polymers. The need extends through all three of the
performance characteristics —mechanical, barrier and compatibility.
All three of those areas are closely tied together by coupled
diffusion-relaxation process considerations once the permeant
enters the polymer. Thus, future researchers should be aware of all
three characteristics when they carry out their particular
experiments in polymer-food interactions and to the extent that is
possible could collect the additional data that will further the
understanding of the relationships between polymer structure and
performance in food packaging applications. Finally, it would be
helpful if publications of food-polymer interactions would include
more detailed data on the composition and properties of the
polymers used. Most useful would be information about the
composition (comonomers, additives) stereoregularity, type and
degree of branching, molecular weight and distribution, and state
of attained crystallinity and morphology, including distribution of

spherulites and defects in the specimens actually used. Similarly, for the food ingredients there is need for more complete description of the chemical composition and treatments to which they were exposed. Such information would be of great use to scientists who wish to help unravel the food-polymer interactions but can not find the observations that would be helpful.

LITERATURE CITED

1. Halek, G. Proceedings of the Fall 1984 Meeting of the R & D Associates for Military Food and Packaging Systems, U.S. Army Natick R & D Center. Activities Report. Vol. 37, No.1, 1985, p. 28-30.
2. Ashley, R. In Polymer Permeability; Comyn, J., Ed.; Elsevier: New York, 1985; Chapter 7.
3. Packaging Encyclopedia 1985; Cahners Publishing: Boston; p. 64 F.
4. Deanin, R. Polymer Structure, Properties of Polymer Films: Cahners Books Boston, 1972.
5. Rogers, C. In Structure and Properties of Polymer Films; Lenz, R. and Stein, R. Eds.; Plenum: New York, 1973; p. 297.
6. Fujita, H. and Kishimoto, A. J. Polymer Sci., (1958), 28:547, 569.
7. Machin, D. and Rogers, C. J. Polymer Sci. A2(1972), 10:887.
8. Nemphos, S., Salame, M., and Steingiser, S. In Kirk-Othmer Encyclopedia of Chemical Technology; Wiley: New York, 1978; Vol. 3, p. 480-502.
9. Combellick, W. In Encyclopedia of Polymer Science and Engineering; Mark, H., Bikales, N., Overberger, C., Menges, G.; Eds.; Wiley: New York, 1985, p. 176-192.
10. Lenz, R. and Stein, S. Eds. Structure and Properties of Polymer Films; Plenum: New York, 1973.
11. Samuels, R. Structured Polymer Properties. Wiley: New York, 1974.
12. Allen, G. and Petrie, S. Physical Structure of the Amorphous State; Marcell Dekker: New York, 1977.
13. Crompton, T. Additive Migration From Plastics Into Food. Pergamon: New York, 1979.
14. Figge, K. In Progress in Polymer Science; Pergamon: New York, 1980; Vol. 6, p. 187-252.
15. Gilbert, S., Miltz, J., Giacin, J. J. Food Processing and Perservation. 1980, 4:27.
16. Niebergall, V., Kutzki, R. Deutsche Lebensmittel-Rundschau. 1982, 78:82.
17. Durr, P., Schobinger, U., Waldvogel, R. Alimenta. 1981, 20:91.
18. Meyers, M., Halek, G. Abstracts of 46th Annual Meeting of Institute of Food Technologists, Dallas, June 1986, Paper 161.
19. Kwapong, O., Hotchkiss, J. Abstracts of 46th Annual Meeting of Institute of Food Technologists, Dallas, June 1986, Paper 162.
20. Shimoda, M., Nitanda, T., Kadota, N. Ohta, H., Suetsuna, K., Osajima, Y. J. Jap. Soc. Food Sci. and Tech. 1984, 31:n.11, 697.

RECEIVED September 24, 1987

Chapter 17

Recent Advances
in Metal Can Interior Coatings

Raymond H. Good

Holden Surface Coatings Limited, Bordesley, Green Road, Birmingham, B9 4TQ, England

Interior Can Coatings have been undergoing a period of considerable change over the last twenty years. Environmental Legislation, new can making technologies and the need to offer increased protection to less corrosion resistant substrates have all influenced coating selection. Tough new air pollution regulations have forced the can producers to adopt compliant water based and high solids coating formulations. The move to eliminate lead from cans has resulted in fundamental changes in can fabrications. The advent of Draw-Redraw (DRD), Drawn and Wall Ironed (DWI) and Welded Technologies all require more highly sophisticated coatings. The coating supplier has responded rapidly to these requirements and continues to work on developing new organic film formers that will withstand the demands of future technologies.

Interior Can Coating Background

Although metal food cans have been with us for 150 years or more, the canning revolution only really started 70 - 80 years ago. In the 19th century, cans were made by hand and a heavy coating of tin protected the foodstuffs from the iron substrate. With the change in lifestyles following the Victorian era, and the greater need for convenience foods, the "tin can" became the necessary package for fruit, meats, fish and vegetables (1). Many of these foods being grown specifically to go into cans.

The improvements in can making and the more widespread use of cans placed an ever increasing interest in reducing the package cost. One of the areas for cost savings was to reduce the tin coating weight and replace it with organic film formers that provide excellent corrosion protection for the metallic container and a barrier between the foodstuff and the metal. Lacquering or coating of the interior of the can was first used to preserve the colour of red fruits and vegetable products when packed in uncoated tinplate

0097–6156/88/0365–0203$06.00/0
© 1988 American Chemical Society

cans. (2). The lacquering of cans was necessary to prevent the blackening of the can and its contents which occurs with certain sulphur products contained in meat, fish and vegetable products. It was Bohart (3) who was active in developing sulphur resistant lacquers in the 1920's. These lacquers contained a suspension of zinc oxide which reacts with sulphur compounds formed during processing to form white zinc sulphide rather than the normal black tin or iron sulphide. (e.g. "corn black").

Most tinplate and all tin free steel containers have to be protected. The exceptions are three piece tinplate cans with heavy tin weights. These are used particularly where the tin aids the flavour and appearance of the product, particularly light coloured fruits, for example pineapple, grapefruit and peaches (4). Protection of a food container has always required the careful selection of lacquer type depending on a number of factors:

Lacquers must have suitable application characteristics either by roll coat, spray, coil, etc. Physically and chemically, lacquers must have excellent adhesion to the substrate, flexibility during forming and durability which provides a film that is odourless and non toxic. Over the years, retention of the flavour and colour of the product has become of paramount importance. This means the lacquer must be stable over a wide temperature range being able to resist the heat from the side seam process and also to the food steam processing conditions.

More recently, coating manufacturers have been restricted in the lacquer choice by components or additives recognised by food legislation,e.g. American Food and Drug Administration.

Chemistry of Interior Can Coatings pre-1950

Oleoresinous Coatings. The original can lacquers were based on oleoresinous products. This group has the longest history in the industry and covers all those coating materials which are made by fusing natural gums and rosins and blending them with drying oils, such as linseed or tung oil. (5) This basic combination explains the term "oleoresinous or oil based" lacquers. The mechanism of drying involves oxidation which is accelerated using driers and temperatures up to 415°F for cure. Tung oil, of which alpha-eleostearic acid is a major constituent, has been used extensively in oleoresinous systems. Its conjugated structure assisting in the oxidative cure (Figure 1).

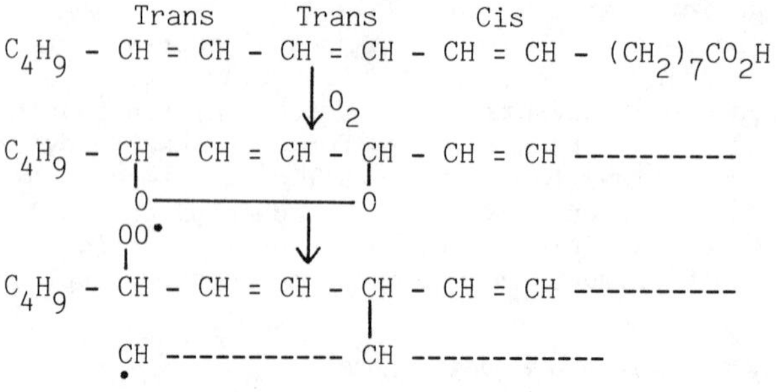

$$
\begin{array}{ccc}
\text{Trans} & \text{Trans} & \text{Cis} \\
\end{array}
$$

$$C_4H_9 - CH = CH - CH = CH - CH = CH - (CH_2)_7CO_2H$$

$$\downarrow O_2$$

$$C_4H_9 - CH - CH = CH - CH - CH = CH ----------$$

$$C_4H_9 - CH - CH = CH - CH - CH = CH ----------$$

$$CH ---------- CH ------------$$

Figure 1 - Alpha-Eleostearic Acid (Tung Oil) Oxidative Cure

The reason oleoresinous systems are still used today is their low cost and good acid resistance (see Table 1). However, because of the open micellar structure of oleoresinous materials, they are prone to corrosion/staining problems with sulphur bearing products unless they are pigmented with zinc oxide paste. It was for this reason that a departure was made from these naturally occuring raw materials to synthetic phenolic resins dissolved in a blend of solvents.

Phenolic Coatings   Phenolic resins are produced by the action of formaldehyde on phenol or other substituted phenols in the presence of a basic catalyst. These phenolic alcohols react immediately with themselves, forming ether bridges or methylene bridges to produce extremely complex molecules which are three dimensional and therefore provide the high chemical resistance (Figure 2). Etherification of the phenolics using a mono alcohol improves solubility in aromatic solvents.

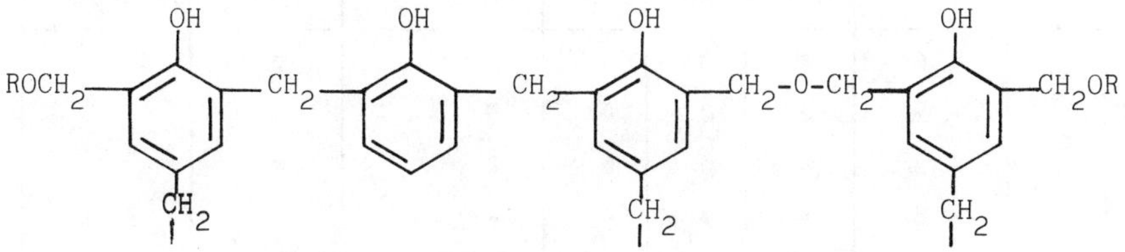

R = H, Alkyl

Figure 2   Possible Resole Phenolic Structure

One of the limiting factors of phenolic coatings is their limited flexibility, high bake requirements and film weight latitude. However, they are still used on three piece bodies where flexibility is not required (see Table 1).

Vinyl Coatings   Vinyl coating materials most relevant to can coatings are based on copolymers of vinyl chloride and vinyl acetate of low molecular weight dissolved in strong ketonic solvents. Minor modification of the vinyl chain with maleic acid enhances the adhesion of the coating to steel and aluminium substrates. The long carbon-carbon chains in vinyl resins make them thermoplastic in nature. Vinyl resins can be blended with alkyd, amino and phenolic resins to enhance their performance (5). See Figure 3 for a typical vinyl co-polymer structure.

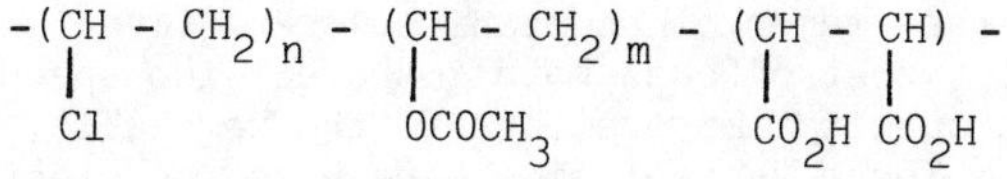

Figure 3 - Typical vinyl co-polymer

The flexibility of these materials allows them to be used for caps and closures and drawn cans. The backbone of a vinyl resin also makes them strong and immune from attack by chemical reagents, such as acids or alkalis. The development of vinyl products has made it possible to develop satisfactory linings for cans to contain beer and soft drinks whose flavour would be affected by the less suitable types of organic film formers. Main disadvantages of vinyl coatings are their sensitivity to heat and also to steam processing of food products. Vinyl coatings were therefore restricted to hot fill and beer and beverage products (Table 1).

Table I - Early Interior Three Piece Can Coatings

	Film Wt g/m²	Int. Bodies	Int. Ends	Meat Fish	Vegetables Soups	Acid Fruits	Beer Beverages
Oleores	4 - 6	●	0	-	-	●	-
Oleores + ZnO	7 - 9	●	0	●	●	-	-
Phenolic	2 - 3	●	0	●	●	●	-
Vinyl	6 - 8	●	●	-	-	-	●
						As Topcoat	

KEY: ● = V. Suitable, 0 = Borderline, - = Not suitable.

1950 - 1986 Can Coating Developments

Three major factors influencing the direction of the major can makers in the last 30 years have been Economics, Environment and Engineering.

The drive towards greater economics forced the can makers into reducing raw material costs. The use of thinner gauge steel with reduction in tin coating weight have placed more emphasis on more sophisticated organic film formers.

The environmental movement in the U.S.A. in the '60's resulted in tough new air pollution regulations being demanded and legislation passed by the government. This led to can makers having to reduce their solvent emissions from their coating lines. Can makers had two choices, either install incinerators or use coatings that complied with the new laws (6). The Food and Drug Administration had the responsibility to regulate the components that can be used in contact with foodstuffs.

The engineers of the major can companies worked feverishly in promoting new technologies that offered both improved line speeds and containers that offered quality improvements to the vegetable, fruit and beverage packers. These technologies were based on electrical resistance welding of tinplate cylinders for 3 piece food cans, Draw Redraw (DRD) for Tin Free Steel (TFS) food cans and the Drawn and Wall Ironed (DWI) process for aluminium or tinplate beverage containers.

It was inevitable that for the new can making technologies to succeed, other organic film formers had to be developed. The most important developments were the solvent based epoxy phenolic, the PVC organosol dispersion and the water borne epoxy-g-acrylic technologies. Each basic technology has its strengths and weaknesses but when modified can produce organic films having superior properties and able to meet the challenges of the new can technologies.

Welded Food Cans

The traditional method of manufacturing 3 piece cans for food and beverage products has been the lead/tin solder process. In recent years, legislation has forced can makers to adopt alternative methods of side seaming to avoid possible lead contamination of food products. The new preferred 3 piece technology being that of electrical resistance welding of tinplate bodies.

The interior body lacquers used on welded cans are a natural extension of food lacquers used on the straight walled 3 piece soldered cans. However, highly refined epoxy phenolic technology has been developed to satisfy the new demanding areas of the welded can. The use of less corrosion resistant tinplate and thinner gauge material with post beading of the can wall to increase strength has virtually eliminated the use of the basic phenolic coating used previously on the soldered can.

Epoxy Phenolic Coatings

Epoxy phenolic coatings are made either by straight blending of a solid epoxy resin with a phenolic resin or are the products of the precondensation of a mixture of two resins in appropriate solvents. During the curing and baking of the coated film, diverse chemical reactions take place which form a three dimensional structure, combining the good adhesion properties of the epoxy resin and the high chemical resistance properties of the phenolic resin. The chemical structure of the epoxy phenolic resin can be represented as shown in Figure 4. Precondensation of the epoxy and phenolic resins leads to better application properties of the coating (7).

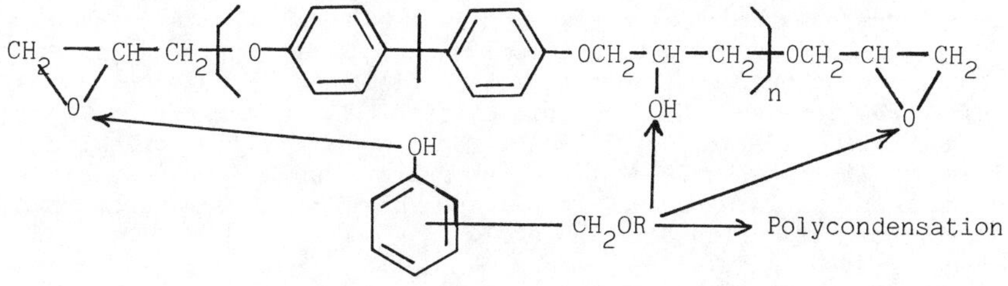

Figure 4 - Epoxy Phenolic Chemistry

The balanced properties of epoxy phenolic coatings have made them almost universal in their application on food cans with the exception of deep multistage DRD cans. Other film formers, based on polyester technology, have made some in-roads into food cans but their hydrolytic instability has made them only applicable for the less corrosive products.

Side Seam Protection: There is an exposed cut edge on the inside of the welded can where most of the tin, during the side seaming process, has been lost. It is necessary to protect this area with a lacquer film with high integrity. Side seam protective coatings have a very important role to play. Without them the foodstuff would be tainted with high dissolved iron and, in extreme cases, perforations would occur.   For aggressive products, the side seam area is typically protected using an electrostatically applied thermoplastic polyester powder (8).   Fusion of the powder results in 100% protection of the weld area. For less aggressive products solution stripes, based on epoxy or vinyl organosol technologies, can be applied using a conventional spray.

Welded can stock will continue to rely on sheet coating because current resistance welding requires a narrow uncoated strip to be left across the width of the coated web. An organic film would interfere with the welding process. Laser welding could utilise plate that has received 100% covering of a coating material.

Draw Redraw Technology

The single shallow drawn cans for fish, meat and petfood products have existed for the last 30 years.

In the 1970's several major can companies decided to invest in the multi-drawn steel container for a whole range of processed foods.

Triple drawn food cans are now a commercial reality. They are lead free. Their integral end and lack of side seam has dramatically reduced the incidence of iron pick up and micro-leakage. All of these problems have for many years been a disadvantage of the 3 piece soldered food can.

In the DRD process, cups are punched from pre-coated tinfree steel, redrawn in two stages, the bottom profile is inserted, cans are trimmed and flanged, beaded and 100% leak tested before being palletised and shipped to the packer.

The preferred base steel, tin free steel known as TFS-CT plate (CT designates chromium type), consists of a very thin layer of chromium metal covered with a chromium oxide film to promote coating adhesion and to prevent undercutting and filiform corrosion.

The DRD process places extreme stretching and compression forces on the upper side wall of the can. Add to this partial thinning of the sidewall by the ironing process and it can be appreciated that coatings for triple drawn food cans need very high performance requirements, as outlined in Table II. (9)

Table II - Coating requirements for triple drawn food cans

0 Good commercial application	0 Hiding power
0 Flexibility and adhesion	0 Colour choice
0 Process resistance	0 Flavour free
0 Chemical resistance	0 FDA approval

Extremely high flexibility and adhesion is needed to resist fracture during the drawing process.

The selection of suitable coatings for this container proved to be a lengthy process requiring extensive test pack evaluations. The interior and exterior coating of choice for triple drawn and partially ironed cans is the PVC vinyl organosol. Extensive studies have shown that certain flexible polyesters, flexible epoxy phenolics and high molecular weight phenoxy coatings may have the flexibility and adhesion to withstand a triple draw but not the properties needed for the extra partial wall ironing process.

PVC Organosol Coatings: The most commonly used interior coating on DRD cans in the U.S.A. is a buff coloured vinyl organosol. A basic PVC organosol formulation will incorporate a high molecular weight PVC organosol dispersion resin which is thermoplastic in nature and extremely flexible. In order to enhance the films product resistance and adhesion, soluble thermosetting resins, including epoxy, phenolic and polyesters are added. Plasticisers are added to aid the film formation. The hybrid system is depicted in **Figure 5**. The cold wet film has the dispersed high molecular weight PVC particles suspended in a blend of aromatic and ketonic solvents. On heating the film, there is a loss of diluent or non solvent which enriches the particles with the stronger ketonic solvents. The continuing action of heat and solvents causes the PVC resin to swell. The plasticiser present assist in the formation of a continuous film as the true solvent is slowly lost. During the latter stages of drying, the soluble thermosetting resins will cure to give the optimum properties needed for the DRD container.

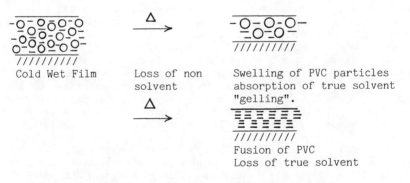

Cold Wet Film   Loss of non   Swelling of PVC particles
                     solvent   absorption of true solvent
                                          "gelling".

                                          Fusion of PVC
                                          Loss of true solvent

Solvents Diluents   eg Xylene - No solvent action
          Dispersants eg Ketones - Solvent action

Figure 5 - Dispersion fusion process

Pigmentation with titanium dioxide is commonly used to produce a white or buff coloured internal coating. Internal lubricants based on high melting waxes facilitate the drawing process (10). DRD cans are now used extensively in the USA and UK for meat, vegetables and pet food.

With the improvements gained by using the 2 piece DRD and welded can technologies, the 3 piece soldered food can is quickly disappearing from the store shelves. By the year 1990 the cans with soldered side seams will be a collectors item.

Beer Beverage Can Developments

It was back in 1935 that the Krueger Brewing Co. in the U.S. first produced a 3 piece soldered tin plate beer can. Since that date, technologies have improved tremendously. Aluminium impact extruded 2 piece cans were tried in the late 1950's. Cemented and welded TFS cans were popular in the late '60's and early '70's. However, from the early days of 1959, developments at the major packaging, aluminium and steel companies ensured the 2 piece Drawn and Wall Ironed (DWI) container has a dominant position today in North America and Europe.

Because of recyclability and deposit legislation, aluminium dominates the 70 billion drinks can market in the U.S. In Europe, DWI was introduced rapidly in the 1970's by the major can companies. Tinplate is used on over 60% of the DWI lines in Europe.

The process: Unlike the DRD process, the DWI manufacturing process begins with a coil of uncoated plate that is unwound, lubricated and fed into a press to form shallow cups. Cups are then fed into an ironing press. The ironer redraws and irons the can walls. During the drawing, ironing and trimming operations the shell is covered with a film of lubricant and coolant. This is removed in the washer where the cans are conveyed, upside down, through a series of cleaning and chemical treating zones with a final deionised water wash. These treatments are necessary to improve coating adhesion. Following the exterior decoration with base coatings, inks and varnishes, the application of the all important interior coating is made by spray. The cans are then baked under strictly controlled conditions. The final stage of metal forming consists of die necking (or spin necking) followed by flanging, testing and palletising.

Spray Coating Chemistry   Traditionally, coatings for beer and beverage cans have been based on a few solvent borne formulations (Table III). Initially, a two coat solvent based system was used on tinplate 2 or 3 piece cans. The basecoat was an epoxy phenolic material giving excellent adhesion and allowing an impermeable vinyl topcoat to be applied. This system is still arguably the best protective system around for beer/beverage containers.

The epoxy/vinyl system has been superseded in Europe by 1 or 2 coats of an epoxy amino clear lacquer system which meets the requirements of the demanding European market for tinplate DWI cans.

In the U.S., with the majority of DWI can lines being aluminium, the lower cost water based interior lacquers now dominate this market.

Table III - <u>Spray Coating Selection for Beer Beverage Cans</u>

1. Solvent based.
     Base coat - Epoxy phenolic
     Top coat   - Solution vinyl.

2. Solvent based
     Epoxy amino, one or two coats.

3. Water based
     Epoxy-g-acrylic copolymer, one or two coats.

    The tremendous growth in the beer and beverage can market in the U.S. and Europe highlighted the real need to reduce pollution from organic solvents. In the U.S., the maximum permitted volatile organic content of spray is 4.2 lbs/gallon calculated with water excluded. In approximate terms, a typical water based spray lacquer contains 20% of film forming resins, 65% water and the remaining 15% is organic co-solvents. To meet the requirements of the EPA, many research departments within the major coating suppliers, spent considerable time and effort in pursuing the need for a water based interior spray material.
    In Table IV the major requirements for a successful water borne interior coating for aluminium or tinplate DWI cans are outlined.

Table IV - <u>Requirements for Water Borne Can Coatings</u>

0   Low applied cost
0   Meet E.P.A. Regulations (<VOC 4.2 lbs/gallon)
0   Acceptable application and complete coverage
0   Good corrosion resistance
0   Impart no off flavour to product
0   Film complies with F.D.A.
0   Good abuse resistance.

Water Borne Chemistry: Given these criteria and utilising the chemistry described previously for solvent based application (Table III), attempts were made to produce equivalent water based formulations. Application, flavour and stability problems arose when pursuing water reducible acrylic, vinyl epoxy phenolic and amino formulations.
    In addition, the design of water borne or water reducible polymer systems without introducing hydrolytically susceptible groups proved difficult. Glidden's approach was to graft ionisable acrylic species onto a hydrophobic epoxy backbone (<u>11-13</u>).

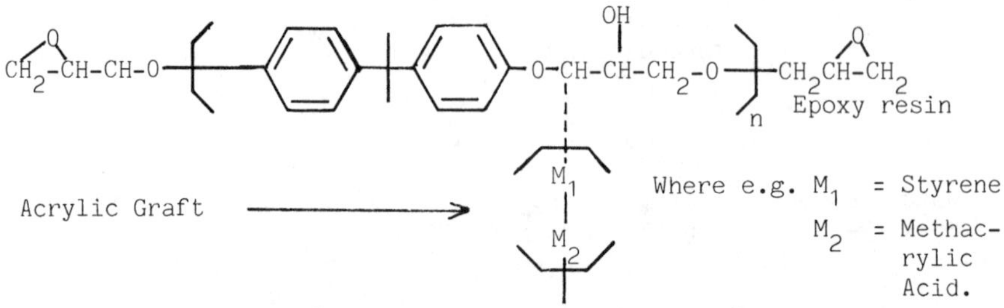

Figure 6 - Epoxy/Acrylic graft co-polymer

   The carboxyl functional resultant polymer was visualised as having a comb-like structure with the grafted acrylic components being of short length, compared to the length of the main polyether backbone (Figure 6).   As with most grafted copolymers, grafting efficiency was not 100% and so the composition of the actual product was a mixture comprising the unmodified epoxy backbone, polyacrylic components and the epoxy-g-acrylic copolymer.   Model compound experiments were carried out and ^{13}C NMR studies have shown most of the grafting takes place on the methylene groups alpha to the aryl ether groups.   The carboxyl acid functional grafted epoxy resin is 70% neutralised with dimethyl ethanolamine and dispersed with deionised water. An aminoplast resin is added to crosslink with the graft co-polymer to produce a final spray formulation having excellent properties.   Coatings produced from such grafted co-polymers have achieved full commercial viability and today are used on more than 70% of the 70 billion beer/beverage cans in the U.S. This technology has particular inertness imparting no undesirable organoleptic property or haze to the canned beverage.

Conclusion: Certain factors are influencing the choice of interior spray lacquers for future beer/beverage cans.   Stricter solvent emission standards are being introduced both in the U.S. and Europe. The further reduction in the neck diameter of cans using a procedure called spin necking highlights the need for greater flexibility from the lacquer.   In addition, more corrosion resistant lacquers are needed to meet the demands for beverage containers with longer shelf lives.   Finally, water based technology is being extended to give protection for the interior of D.W.I. food cans.

Future of the Lacquered Food Can

The future of the lacquered metal container for food and beverage is under attack from many areas of rigid and plastic packaging. However, the metal container still remains the ultimate barrier for processed foods.
   The can makers are determined the metal can is here to stay and are investing heavily in reducing costs, increasing line speeds and improving quality.   A key to the success in these areas is that of lacquer development which continues at a fast pace.

Areas in which coating companies will be directing their efforts to meet the requirement of future metal packaging developments are listed in Table V.

Table V - Future Areas for Can Coating Research

Higher Solids
- Dispersion Chemistry
- Two Pack Systems
- Powder Coatings
- UV/electron beam curable

Water Based
-     100% W/B Technology
- Anodic/Cathodic
  Electrodeposition

Higher solids solvent based materials could be based on novel dispersion chemistry. Two pack coatings will give low temperature cure and thereby saving energy. 100% Powder coatings are obviously the ultimate environmentally because of the lack of solvents. Current problems however have been that the particle size of the powder has not been reduced sufficiently so that a 100% continuous film can be achieved economically. UV curable or electronbeam curable systems could be designed for the interiors of cans. Currently, U.V. curable coatings are restricted to the exterior of metal cans because of the non-F.D.A. approval of these materials. Novel water based technologies are being researched to provide coatings that require less organic cosolvent for application and flow. The anodic electrodeposition of can coatings is reaching an advanced commercial stage and is worthy of more detailed discussion.

Electrocoating of Food Cans

The electrocoating of cans and can stock provides a unique technology for protecting both steel and aluminum. It is possible that this technology could revolutionize can coating as we know it today.
As a result of a joint development between Metal Box and Alcoa, this year could see the start up of the first commercial electrocoat line designed to 100% coat a fully formed drawn and redrawn aluminium food can.

The Process   Triple drawn cans are produced from uncoated aluminium coil using excess lubricant. Before being electrocoated each can is tested for pinholes, washed to remove excess lubricants and treated and finally washed with D.I. water. On the coating carousel, can bodies are fed into an open cell (**Figure 7**).

The cell then closes around the can and coatings are applied to both the inside and outside can surfaces simultaneously. An electric potential is applied between the mandrel (the cathode) and the can body (anode) which lasts less than 1 second. The charged resin particles are attracted to the can body forming a very uniform film over the container surface. As the coating deposits the film weight reaches a certain thickness and electrical resistance increases in these areas,forcing the charge particles to find areas of lower resistance. The whole container is therefore coated very uniformly depending on the current flow within the cell. Different coating weights can be applied to inside and outside by varying the current. The wet or "green" film is made up of coating particles with most of

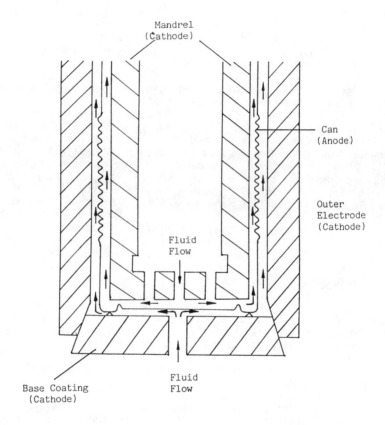

Figure 7 – <u>Simplified Flow Diagram through the Metal Box</u>
<u>MB 180 Electrocoat Cell</u>

the water having been forced out. Excess film former is removed from the container by washing before the can is stoved. The excess coating is then returned to the coating bath and concentrated using ultrafiltration.

Chemistry of Electrocoating science: electrochemistry and polymer science. In anodic electrodeposition, the metal substrate is immersed in an aqueous solution of polyelectrolyte. Current flows through the system causing a high concentration of hydrogen ions near the anode or can wall. The resultant coagulation of acid functional resin particles forms a strongly adhering wet film to the substrate (**Figure 8**). The acid soluble polymers could be based on sophisticated extensions of epoxy or acrylic technologies used for conventional spray application.

At the anode

$$2H_2O \longrightarrow 4H^+ + O_2 + 4e-$$

RCOO— + H^+ $\longrightarrow$ RCO$_2$H   R = Epoxy or Acrylic
Soluble   Insoluble   Polymers

Figure 8 - Basis of anodic electrodeposition

The many advantages of electrocoatings are summarized in table VI.

Table VI - Advantages of Electrocoating of Cans

0	High coating utilisation	0 Uniform coating thickness
0	Ability to cover difficult profiles	0 100% coverage of inside and outside
0	Compliant coatings	0 Passivation of aluminium cans
0	High quality	
0	Universal coating possibility	

The process overcomes the current wasteful spray techniques with almost 99% coating utilisation achieved (14). Uniform application over awkwardly shaped profiles can be achieved at speed. Simultaneous coverage of both inside and outside is achieved in less than 1 second. The use of water based coatings low in volatile organic solvents easily meets the EPA regulations. Finally, the aluminium surface is passivated during the E-Coating operation which helps adhesion and improves long term quality of the package.

The electrocoat process will eventually be extended for use on steel food and beverage cans where coatings savings and superior coverage are expected for the more aggressive products.

It can be seen that this process meets the requirements of the environment while producing coated cans at speed with excellent quality. It is an example of what research and development can bring to the packaging industries and could at last herald the discovery of the elusive universal coating for both food and beverage cans.

Conclusion

The transformation of can coatings from traditional to high technology products is accelerating as can making changes. Close collaboration between coating research departments and the can makers will ensure a healthy future for metal packaging.

Acknowledgments

Metal Box P.L.C., England
The Glidden Company, Ohio, U.S.A.
Holden Surface Coatings Limited, England.
Holden Europe S.A., France.
I.C.I. Lacke Farben, Wiederhold, West Germany.

Literature Cited

1. Reader, W.J. Metal Box - A History; Heineman : London, 1976 Chapter 1.
2. May, E.C. The Canning Clan; Macmillan Co. : New York, 1938.
3. Bohart, G.S. National Canners Association Circ. 10L, 1924.
4. Howard, A.J. Canning Technology; J & A Churchill Ltd. : London, 1949. Chapter 4.
5. Pilley, K.P. Lacquers, Varnishes and Coatings for Food and Drink Cans; Holden Publication, 1968, p.40.
6. Guerrier, J. Second International Tinplate Conference; I.T.R.I. Plublication No. 600, 1980, P.434.
7. Guerrier, J. La Corrosion des Boites Metalliques destinees a l'Industrie Alimentaire; ASBL-INACOL, 1971, p.59.
8. Falkenburg, H.R.; McGuiness R.C. Second International Tinplate Conference; I.T.R.I. Publication No. 600, 1980, p.442.
9. Good, R.H. Presentation at NCCA Conference, Weirton, July 26, 1984.
10. Carbo, A; Good, R.H. U.S. Patent 4556498, 1986.
11. Robinson, P.V. Journal of Coatings Technology; 1981, 53, 23-30.
12. Evans, J.M.; Ting, V.W. U.S. Patent 4212781, 1980.
13. Woo, J.T.K. et al ACS Symp. Ser. 221; Epoxy Resin Chem 2, 1983, 284-400.
14. Clough, A. Packaging Week; 1987, 2 (34), p.1.

RECEIVED November 5, 1987

Chapter 18

Oxygen and Water Vapor Transport Through Polymeric Film

A Review of Modeling Approaches

Roy R. Chao and Syed S. H. Rizvi

Institute of Food Science, Food Science Department, Cornell University, Stocking Hall, Ithaca, NY 14853-7201

The dramatic growth in the use of polymeric films for food packaging asserts their many inherent advantages over other materials. However, during storage, such undesirable transport phenomena as permeation of moisture, oxygen and organic vapors through the polymeric film do occur and their knowledge and control become critical. These transport processes are affected by the thermodynamic compatibility between the polymer and the penetrant and the structural and morphological characteristics of the polymeric material. In this review, some modeling approaches for describing the transport of oxygen and/or water vapor through hydrophobic or hydrophilic polymeric films in terms of variables generally evaluated for the film and the environmental conditions established within the package during storage are discussed and analyzed.

Prolonging the consumer-acceptable shelf life of food products is the major benefit bestowed by the use of packaging. The first step in defining packaging requirements is quantifying the transport properties and the critical vectors of quality loss as well as the variables that influence them [1]. The importance of transport behavior of gases and vapors in polymeric films has become apparent with the accelerating development of highly impermeable or selectively permeable packaging films for diverse applications in the food and pharmaceutical industries.

Usually, films or coatings used for packaging foods are expected to resist transfer of moisture and non-condensable gases like O_2 and CO_2, sorption of fats/oils and migration of additives.

In some instances, however, regulated transfer of some of them may
be desirable. Other required functional properties of films
include: protection against mechanical hazards during
transportation, structural and sanitary integrity, retention of
volatile flavor compounds and enhancement of sales appeal (2-3).
Prediction of shelf life, maximization of economic benefit and
regulatory considerations generally dictate the properties of
packaging materials, and transport properties in particular, must
be specified. Therefore, development of models describing
accurately the specific transport of gases and vapors through the
films would be highly desirable. Many extensive reviews
pertaining to modeling of permeation of gases and vapors in
polymeric materials have been reported in the literature (4-20);
and over 7,000 apparently relevant and peripheral papers in the
past two decades are available in the Chemical Abstract database
(21). This review is, therefore, concerned only with aspects of
the solution and diffusion of oxygen and water vapor through
polymeric materials. Attention has been directed specifically at
fundamental understanding of the processes and approaches taken to
quantify them.

General Theory of Permeation

There are usually two types of mechanisms of mass transfer for
gases and vapors permeating through packaging materials; namely, a
capillary flow type and an activated diffusion type (22).
Capillary flow involves small molecules permeating through
pinholes and/or highly porous media such as paper, glassine,
cellulosic membranes, etc. Activated diffusion consists of
solubilization of the penetrants into an effectively non-porous
film at the inflow (upstream) surface, diffusion through the film
under a concentration gradient and release from the outflow
(downstream) surface at the lower concentration.

   In flexible packages of either highly porous media or with
gross defects such as cracks, folds, and pinholes, the dominant
mechanism is capillary flow which actually determines the
transmission rates. The flow rate of a penetrant in such cases
depends on the size of the capillaries, size and viscosity of the
penetrant, and both the total pressure and pressure differential
across the film. In the case of non-porous polymeric films, the
mass transport of a penetrant basically includes three steps:
adsorption, diffusion, and desorption. Adsorption and desorption
depend upon the solubility of the penetrant in the film, i.e. the
thermodynamic compatibility between the polymer and the penetrant.
Diffusion is the process by which mass is transported from one
part of a system to another as a result of random molecular
motions. Within the polymer matrix, the process is viewed as a
series of activated jumps from one vaguely defined "cavity" to
another. Qualitatively, the diffusion rate increases with the
increase of the number or size of cavities caused by the presence
of substances such as plasticizers. On the other hand, structural
entities such as crosslinks or degree of crystallinity decrease
the size or number of cavities and thereby decrease the diffusion
rate.

The migration of the penetrant in a polymeric film can also be visualized as a sequence of unit diffusion steps or jumps during which the particle passes over a potential barrier separating one position from the next. The diffusion process requires a localization of energy to be available to the diffusing molecule and its polymer chain segment neighbors. Theoretically, the localization of energy provides what is needed for rearrangement against the cohesive forces of the polymeric medium with effective movement of the penetrant for a successful jump. A more thorough discussion on the diffusion process is provided by Roger (21).

Permeability and Related Equations

At constant temperature and differential partial pressure, diffusion of a penetrant through a polymeric film of unit area normal to the direction of flow for a period of time leads to a steady state of diffusive flux, J,

$$J = Q/At \tag{1}$$

where Q is the total amount of penetrant which has passed through surface area A in time t. Fick's first law of diffusion also applies:

$$J = -D(\delta c/\delta x) \tag{2}$$

where x is the space coordinate normal to the reference place; c is the concentration of the diffusing penetrant; and D, is the diffusivity, assumed independent of x, t, or c.

For experimental and predictive purposes, the diffusing concentration, c, is usually related to the ambient penetrant concentration, C, in contact with the polymer surface by the Nernst distribution function

$$c = KC \tag{3}$$

where K is the distribution coefficient and is a function of temperature. Often, penetrant concentration, c, is also proportional to pressure, p, through an appropriate gas law equation. For example, when Henry's law is obeyed ($c = pS$), it follows that the steady state flux can be written as

$$J = DS(p_1 - p_2)/L \tag{4}$$

where p_1 and p_2 are the respective upstream and downstream pressure of the film of thickness L, and S is the Henry's law solubility coefficient, assumed independent of p and c.

By defining the permeability constant, P_m, as the product of solubility and diffusivity, Equation 4 is used for evaluating the permeability constant,

$$P_m = DS = \frac{L \cdot Q}{A \cdot t} \cdot \frac{1}{(p_1 - p_2)} \tag{5}$$

This is the basic equation for determining the permeability constant. However, D and S generally vary with c, p, x, or, t so that P_m will also be dependent on those variables.

Experimental methods and apparatus for the measurement of S, D, and P_m have been described by many investigators. Current commercial gas and vapor transmission instruments include the ASTM (American Society for Testing Materials) standard test Dow-Park cell, the Linde cell, cells for use with a mass spectrometer detector, various water vapor transmission cells and the Mocon instruments for CO_2, O_2, and water (19). These instruments are adequate for the determination of permeability at steady-state. Commercially available microbalances are used to measure rates of sorption and desorption. Other systems using volumetric, optical radiotracer, weighing cup and a variety of other measurement methods has also been described (19).

Unsteady State Diffusion Measurements

When a penetrant diffuses through a polymeric film in which it is soluble, there is a transient state before the steady state is established. Two approaches are used to quantify the unsteady state process.

Integral Permeation Method In this technique, the penetrant pressure p_1 at the upstream face of the polymer is generally held constant. Meanwhile, the downstream pressure, p_2, while measurable, is negligible relative to the upstream pressure. When Henry's law is applicable, a typical plot of gas transmission versus time appears as shown in Figure 1. The intercept on the time axis of the extrapolated linear steady state portion of the curve, τ, is known as the "time lag" and can be expressed as (23)

$$\tau = L^2/6D \tag{6}$$

Therefore, all three parameters – P_m from the steady state portion of the curve, D from the time lag, and S from P_m/D – can be determined from the single experiment by analyzing steady and non-steady flow through a membrane (24). Solubility constants determined by the time-lag method, however, are less precise for more soluble gases, e.g. carbon dioxide, ethane, etc., than those determined by the equilibrium-sorption method. The precision limits on solubility constant by the time-lag method are estimated at $10 \pm 1.0\%$ (25).

For a penetrant-independent diffusivity, the steady-state flow through a plane sheet is approximately reached after a period amounting to 2.7 τ (16).

If the diffusivity, D, is not constant, but depends upon c, then a mean value of the diffusivity, $\underline{D}$, may be defined as

$$\underline{D} = (1/C_1) \int_0^{C_1} D(c)d \qquad (7)$$

where C_1 is the penetrant concentration at the upstream penetrant membrane surface. Generally, as long as the upstream concentration does not exceed roughly 10 wt. percent of the polymer, the diffusivity can be described by the following equation

$$D(C) = D_0 \exp(Gc) \qquad (8)$$

where D_0 is the value of $D(c)$ at zero diffusing component concentration; G is a "plasticizing parameter" that can be related to the Flory-Huggins polymer-solvent interaction parameter (12,26).

It has been shown (27) that the following inequality holds for a large class of fundamental dependencies of $\underline{D}$ on penetrant concentration

$$1/6 <\underline{D}r/L^2 < 1/2 \qquad (9)$$

This relation indicates that using Equation 6 to estimate $\underline{D}$ may lead to the estimated value being up to three times as small as it should be. Although a convenient and relatively simple means of evaluating the transport parameters of gases in polymeric films, the time-lag method has some drawbacks which limits its effectiveness. One major drawback is that, in Equation 6, the concentration of penetrant in the receiver is assumed to be negligible compared to that in the reservoir. Correction for pressure buildup in the receiver is possible, but additional errors in the computation of D will result.

Diferential Permeation Methods The differential permeation method has been used by Ziegel et al. (28) and Felder (29) to improve the previously mentioned major drawback. In this technique, the membrane serves as a barrier between two flowing gas streams, penetrant and carrier, each at atmospheric pressure. The permeation rate is obtained by analyzing the downstream effluent gas to determine the penetrant concentration, and multiplying the concentration by the gas flow rate.

Data of a typical continuous-flow permeation system are shown in Figure 2. The permeation rate, ϕ, is measured as a function of time, till the steady state permeation rate, ϕ_∞, is obtained. For a flat membrane, ϕ_∞ given as

$$\phi_\infty = (P_mA/L)(p_1 - p_2) \qquad (10)$$

Four techniques have been used for estimating the diffusivity from a curve of the form shown in Figure 2. The first estimation

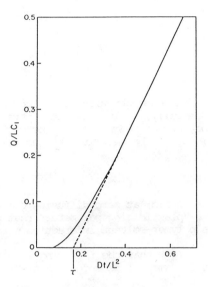

Figure 1. Approach to steady-state flow through a phase sheet. (Reproduced with permission from Ref. 16. Copyright 1975 Oxford University Press.)

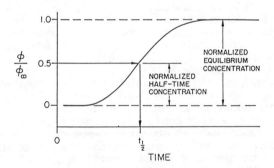

Figure 2. Use of half-time techniques for calculating diffusivity as in equations 10 and 11.

technique, the so-called half-time method, is to designate the time $t_{1/2}$ at which ϕ is equal to 0.5 ϕ_∞. For a flat membrane this becomes (27)

$$D = \frac{L^2}{7.199 \; t_{1/2}}$$ (11)

The second estimation technique is similar to a treatment given by Jost (30) for diffusion into or out of a thin slab

$$\ln(1-\phi/\phi_\infty) = \ln 2 + \ln \sum_{n=1}^{\infty} -(-1)^n \exp\{-n^2\pi^2 Dt/L^2\}$$ (12)

As time becomes sufficiently large, all terms other than n = 1 in the series vanish; hence, the linear portion of a plot of $\ln(1-\phi/\phi_\infty)$ versus t has an intercept of ln2 at t = 0 and a slope related to the diffusivity by

$$D = -L^2(\text{slope})/\pi^2$$ (13)

The third estimation method used by Rogers et al. (31), is based on the Holstein relation (16) for small time as shown below

$$\ln(\phi t_{1/2}) = \ln[2c(D/\pi)^{1/2}] - L^2/4Dt$$ (14)

The slope of a plot of $\ln(\phi t_{1/2})$ versus 1/t yields $(-L^2/4D)$ from which D can be calculated. The advantage of this method is that it does not require knowledge of the value of ϕ_∞. However, as shown by Yasuda and Rosengren (32), the values of D estimated in this manner seem to be more sensitive to the gas flow rate as compared to the values obtained by the half-time method.

Pasternak et al. (33) developed several other small-time estimation techniques for flat membranes, and, although their dynamic approach to diffusion and permeation measurement is particularly suited to the study of transient diffusion in polymers, their mathematical solutions still require a knowledge of ϕ. The method used is valid over a wider range of times than that used by Rogers et al. (31).

The fourth estimation method is a moment technique used by Felder et al. (29,34,35) who defined a quantity, t_p, equivalent to the time lag of the closed volume experiment.

$$t_p = \int_0^{\infty} [1 - \frac{\phi(t)}{\phi_\infty}] \, dt$$ (15)

Once t_p is determined, the diffusivity for a flat membrane may be estimated as

$$D = L^2/6t_p \qquad (16)$$

The moment method is claimed to have several advantages (19): (i) it requires simple numerical integration rather than curve fitting; (ii) it utilizes the complete response rather than a single point (as in the half-time method) or a portion of the response that falls within the region of validity of a small-time asymptotic solution of the diffusion equation; and (iii) it enables the dynamics of system components rather than the membrane to be factored out of the measured response.

A more fundamental and comprehensive approach to diffusion may rely on the theory of irreversible thermodynamics (36–38). A rigorous development on that basis is beyond the scope of this review.

Transport Models Related to Oxygen and Water Vapors

Modeling approaches for evaluating transport of penetrant through films may be mechanistic, purely empirical or phenomenological, or intermediate between mechanistic and empirical (39). Mechanistic models are typically used to provide methods for determining how transport is affected by the film structure and composition and by physicochemical interactions between penetrants and the film (40–42). For example, in a semi-crystalline polymer, the permeability of the crystalline phase is slow enough to be neglected. The overall permeability constant for semi-crystalline polymer may then be expressed as (43–45)

$$P_m = X_a S_a D_a/hb \qquad (17)$$

where X_a, S_a, and D_a are the respective fraction, solubility, and diffusivity of the amorphous phase; h is a geometric factor connected with the length and cross-section of the diffusion trace; and b is the mobility of chain segments associated with the crystalline linkage.

Some mechanistic models, when adequately tested and verified, can provide reliable prediction of permeation behavior. Many other mechanistic hypotheses, however, may not be valid over a reasonably wide range of conditions (39). Purely empirical models have been extensively used by many investigators, but the models tend to be poorly defined and the results deduced from such models are likely to be in doubt. Intermediate models are, therefore, frequently applied in which well-defined portion of the transport process are described as accurately as possible, and then some suitable formal models, for example, Fick's laws of diffusion, are used to describe the in-film transport of penetrants (39). It should be born in mind, however, that suitable formal models may not always be valid in a mechanistic sense. Also, the success of models to represent transport data of penetrants does not necessarily prove the validity of the specific model.

Modeling for transport of penetrants usually depends upon the thermodynamic characteristics of polymeric films, especially whether they are hydrophobic or hydrophilic. In the case of flexible films for packaging dehydrated foods, polymers with low thermodynamic compatibility with water (hydrophobic) are preferable. Solubility of water in the film is thus very small and water vapor permeation is minimal. Hydrophilic films, on the contrary, are thermodynamically compatible with water and infusion of water leads to swelling of the film. Such barrier is observed in some vinyl and cellulose types of packaging materials. In this case, the solubility of water in the film is not negligible, and diffusion rate for all diffusing species are usually strongly affected by the water content of the film.

Permeation of Water Vapor in Hydrophobic Polymers

Recognizing the sensitivity of both the packaged dehydrated food and the hydrophobic packaging materials to vapor sorption, Peppas and Khanna (46) developed a model to account for these two related phenomena. In developing the model, they started with the general Nernst-Planck's diffusion equation applicable to the transport of n penetrants through polymeric films (47,48):

$$N_i = -CD_{im}d_i + X_i \sum_{\substack{j=1 \\ j \neq i}}^{n} (N_j/D_{ij}) \Big/ \sum_{\substack{j=1 \\ j \neq i}}^{n} (X_j/D_{ij}) \tag{18}$$

where N_i is the total mole flux of species i; X_i is the mole fraction of species i; C is the total concentration; D_{ij} is the multicomponent diffusivity for the pair ij; D_{im} is the effective binary diffusivity for the diffusion of i in a mixture; and d_i is the driving force for transport of the ith component through the system. With the following assumption of (46):
- Homogeneous polymeric film,
- Only water vapor diffusing through film,
- Henry's law dissolution constant applicable to link the in-film water concentration with the external water vapor pressure,
- Small solubility of the water vapor in the film,
- Sealed package without external/internal pressure difference,

Equation 18 is modified into a simplified equation for water vapor transport

$$\frac{dm}{dt} = \frac{P_w S}{m_s} (a_w^e - a_w^i) \tag{19}$$

where m is mass of water adsorbed per mass of food product; m_s is mass of packaged food product P_w is the water vapor permeability;

a_w^e and a_w^i are the respective external and internal water activities. Values of the water vapor permeability, P_W, of selected commonly used polymeric films were listed elsewhere (49).

Equation 19 is actually another integrated expression of Fick's law written in terms of water activity which is widely used in the food packaging area. Under constant atmospheric conditions, a final value of water activity inside the package will be attained as a result of the sorption characteristics of water on the packaged food product. Most food sorption isotherms can be adequately described by one or more of the following equations, e.g., linear, Langmuir, Brunauer-Emmett-Teller (BET), Halsey, Oswin, Freundlich, etc. In recent years, the Guggenheim-Anderson-de Boer isotherm equation has been shown to better describe the sorption behavior of most foods from 0.1 to 0.9 a_w at temperature from 25 to 65°C (50-54).

As pointed out by Peppas and Khanna (46), when a single isotherm equation is not adequate to describe the sorption characteristics of certain food systems over a wide range of water activity, a combined sorptive mechanism may be imperative. For example, during the initial period of storage, the BET isotherm may be suitable for lower water activities, while the Oswin isotherm may be applicable at higher water activities. Equation 19 may conjugate with one of the sorption isotherms to obtain a specific solution of the resulting differential equations in terms of internal water activity and adsorbed water as a function of time. Further, any specific solution due to different isotherm equation always leads to a so-called permeability-sorption constant, P_{ms} (day^{-1}), written as

$$P_{ms} = M \cdot F_S \tag{20}$$

where $M = P_W S/m_S$, is the diffusive contribution term, and F_S is the sorption contribution form. Detailed forms of F_S have been covered elsewhere (46).

To illustrate the applicability of the model, Figure 3 shows a typical prediction of the internal water activity of dehydrated apples packaged in four typical polymeric films, e.g., polyethylene (PE) coated cellophane, polyvinylidene chloride (PVDC) coated cellophane, PE, and plasticized PVDC films. The curves in Figure 3 were obtained (46) using experimental data reported by Rotstein and Cornish (55), and data on water vapor permeability were from Karel et al. (56). It is seen from Figure 3 that the plasticized PVDC film (curve 4) provides the best protection to dehydrated apple during storage. However, if a storage time of only 70 days and a final water activity of 0.3% are desired, then PE film (curve 3) would be adequate as a packaging material.

Oxygen Transport through Hydrophobic Film

The modeling approach developed by Peppas and Khanna (46) deals only with the transport of water through hydrophobic films.

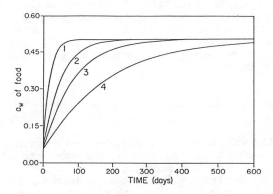

Figure 3. Prediction of the internal water activity in packages sealed at a_w = 0.05 and stored at w_w = 0.50 using various polymeric films. Curve 1: PE coated cellophane; 2: PVDC coated cellophane; 3: PE; 4: plasticized PVDC. (Reproduced with permission from Ref. 46. Copyright 1980 Society of Plastics Engineers.)

Similar approach can also be used to describe the permeation of oxygen through film for oxidation sensitive foods.

Oxidative degradation (57-60) involves oxygen diffusion through the film, sorption and simultaneous reaction on the food surface. For oxygen diffusion, Khanna and Pappas (61) transform Equation 19 to

$$\frac{dV}{dt} = \frac{Po2A}{L} (P_e - P_t) \qquad (21)$$

where V is the volume of diffused oxygen; P_{O2} is the oxygen permeability; and p_e and p_t are the partial pressure of oxygen at the external and internal conditions at time t, respectively. They then used a simple Langmuir kinetic expression, developed by Simon (62) and shown below, to describe adsorption and reaction of oxygen on the surface

$$\frac{dV_r}{dt} = \frac{m_s p_t}{k_{s1} p_e + k_{s2}} \qquad (22)$$

where V_r is the volume of oxygen adsorbed and reacted on the surface at STP conditions; k_{s1} and k_{s2} are constants characteristic of thte food product.

It can be seen that p_t is involved in both Equations 21 and 22 and can be determined by using the following relationship

$$P_t = P_i + P_{T,t} [(V - V_r)/V_u] \qquad (23)$$

where p_i is the initial partial pressure of oxygen; $p_{T,t}$ is the constant total pressure inside the package; and V_u is the unfilled volume of the package. An analytical solution can then be derived to determine p_t by diffrentiating Equation 23 and substituting into Equation 21 and 22. The solution is applicable in the range of 0.35 and 0.60 a_w (63). In this range, changes in water activity have negligible effect on the rate of oxidation. Once a set of packaging parameters has been determined, the shelf-life of the packaged food can then be predicted, provided that the maximum allowable concentration of oxygen in the package is known.

To illustrate applications of the theoretical analysis to specific packaging conditions, packaging parameters of a packaged shrimp system for the study of carotenoid loss (62) were applied to thet analytical solution, and the partial pressure of oxygen inside package as a function of time was plotted (61). They found that the unfilled volume, V_u, and the value of oxygen permeability, P_{o2}, were very sentive to the storage stability of the packaged shrimp. On the basis of their results, they recommended that a minimum V_u should be used in packaging of oxygen-sensitive foods, and showed that an increase in the value of oxygen permeability by only an order of magnitude could drastically reduce the shelf-life, under the same constant conditions of packaging.

Models for Permeation of Water Vapor and Gases through Hydrophilic
Films

As mentioned earlier, thermodynamic interaction between the vapor
and the polymeric film may lead to considerable swelling of the
packaging material. If rapid equilibrium swelling is reached,
further diffusion of water vapor is done through the swollen
polymer film with constant volume ratio, R_V, where R_V is defined
as the volume of the water swollen film over the volume of dry
polymer (46). In some cases, swelling is time-dependent and, for
instance, a linear relation of swelling with time can be simply
expressed as

$$R_V = R_V^0 + c_i t \qquad (24)$$

where R_V^0 is the initial swelling ratio and equals to one for fully
hydrophobic films. In the case of a semicrystalline crosslinked
polyvinyl alcohol film, where $c_i = 0.023 \ day^{-1}$, this linear
relationship can be maintained for up to 110 days (46). Yasuda
and Lamaze (64) have defined the degree of hydration, H, as the
ratio of weight of in-film water to the total weight of the film,
and can be expressed as

$$\frac{1}{H} = 1 + \frac{1}{(R_V-1)} \cdot \frac{\rho_p}{\rho_W} \qquad (25)$$

where ρ_p and ρ_W are the densities of polymer and water
respectively. They developed a theory for predicting the effect
of degree of hydration, H, on the permeability, P_W, of a solute
through a swollen polymeric film as

$$P_W = \beta H exp[- \gamma (\frac{1}{H} - 1)] \qquad (26)$$

where β and γ are constants depending on the solute size, polymer
structure and temperature.
   Peppas and Khanna (46) utilized the concepts of degree of
hydration and volume ratio to define a new permeability-sorption
constant for water vapor diffusion through polymeric films under
swollen conditions. However, no experimental verification was
provided due to lack of appropriate experimental data in the
published literature with respect to the actual prediction of the
shelf life of food products packaged in cellulose based materials.
   Another modeling approach for water vapor and oxygen
transport through hydrophilic films has been reported by
Schwartzberg (39). In the development of this model, he starts
with a generalized diffusive flux of water, J_W, passing from
upstream surface, 1, to downstream surface, 2, of a film

$$J_w = K_{w,j} \ (p'_{w,j} - p_{w,j})$$
$$= K_{w,j} \ (p'_{w,j} - p^* a_{w,j}) \tag{27}$$

where $K_{w,j}$ are mass transfer coefficients (K_g water/$m^2 \cdot atm \cdot s$); $p_{w,j}$ is partial pressures in atmospheres; p^* is the vapor pressure of pure water; $a_{w,j}$ are the water activities in equilibrium with the respective film moisture contents at surface 1 and 2 respectively; the subscript $'$ is to denote the bulk stream adjacent to both surfaces; j is to denote either side 1 or 2.

Referring to both surfaces of the film, the assumptions are:
- Sorption and desorption occurs quickly and equilibrium is always maintained,
- Heat of sorption equals heat of desorption, and
- Hysteresis with respect to sorption equilibrium can be neglected.

A general expression of the sorption isotherm can be written as

$$X_j = f(a_{w,j}) \tag{28}$$

where X is the moisture content (mass water/mass dry solid) and is a function of water activity.

Referring to the in-film diffusion of water, the following assumptions prevail:
- Steady-state or quasi steady-state exists,
- Free-volume model for diffusivities in polymeric film is applicable.

Then, Fick's first law can be applied and expressed as

$$J_w = -D_w(dX/ds) \tag{29}$$

where D_w is the diffusivity of water in the film and s is the mass of dry polymeric film per unit area between surface 1 and to the level where the moisture content is X. Equation 29 is not written as a usual form of Fick's first law (in terms of unit volume concentrations and geometric lengths). The advantage of the formulation of Equation 29 is the avoidance of computational difficulties imposed due to significant swelling of hydrophilic films upon absorbing a large quantity of water (39). An empirical relationship similar to Equation 8 such as

$$D_w = D_w^o \ \exp(GX) \tag{30}$$

is used to correlate D_w in terms of X. D_w^o is the diffusivity of water in the film at zero moisture content and G is a plasticizing constant. Typical values of D_w^o and G are reported around 2.3 x 10^{-5} (kg solid)^2/($m^4 s$) and 2.9 respectively (39). Applying Equation 27 to Equation 26 leads to

$$J_W = (D_W^O/GU)[\exp(GX_1) - \exp(GX_2)] \tag{31}$$

where U is the total mass of dry polymeric film per unit area of film.

It may be recalled from Equation 4 that J_W is needed to determine the water vapor permeability. Apparently, Equations 27, 28 and 31 are needed to be simultaneously solved in order to determine J_W under appropriate conditions. Often, the interfacial partial water vapor pressures, $p_{W,j}$, usually are unknown and a trial and error type of calculation is needed for the three equations mentioned above to determine J_W. A flow diagram of the computational scheme for solving the three equations has been given (39).

Gas Transport   If Fick's first law is also applied to the transport of condensable gaseous species i across the hydrophilic film, then

$$J_i = -D_i(dC_i/ds) \tag{32}$$

where C_i is the in-film concentration of gas i at trace levels relative to that of water. D_i, in turn, is correlated to the water concentration written as

$$D_i = D_i^O \exp(qX) \tag{33}$$

where q is the plasticizing constant for gases. For oxygen, q is around 6.2 (39).

For hydrophilic films D_i^O is much smaller than D_W^O. Some values of the ratio D_i^O/D_W^O can be estimated based on the data of Kokes and Long (65), Hauser and McLaren (66), and Karel et al. (67).

The flux J_i can be written in terms of J_W and X_2 as

$$J_i = J_W (C_1 - C_2)(q - G)(D_i^O/D_W^O)[\exp(GX_2)]^{(q/G)-1} \tag{34}$$

Equation 34 shows that J_i will be proportional to J_W, but it will be much smaller than J_W due to the small value of (D_i^O/D_W^O). Since a typical value of G is 2.9 and q value for oxygen is 6.2, the exponential term, $(q/G)-1$, of Equation 34 is greater than 1.0. Therefore, J_i increases markedly as X_2 increases. This means that the gas permeation rate tends to increase as the humidity and moisture content on the downstream side of hydrophilic packaging film increases.

The variables involved in the above-mentioned equations to determine J_W and J_i, are $K_{W,j}$, D_W^O, G, U, D, and $P_{W,j}'$. These variables can be determined by conducting standard water vapor permeation tests such as those described by ASTM standard E96-80 (68). Once $K_{W,j}$ and $a_{W,j}$ of Equation 27 are determined, the flux equation of water can then be obtained. The flux of water is subsequently subjected to a suitable sorption isotherm equation, i.e. Equation 28, to determine specific X_j. Through a series of tests carried out using different values of $P_{W,2}'$, a set of J_W and

X_j data are established. This data is then used along with
Equation 29 and a non-linear regression routine (69), to determine
best fit values for the parameters D_w^o and G. A detailed method of
using a non-linear regression routine in conjunction with all
equations and related operation procedures was illustrated by Chao
(70).

The model developed by Schwartzberg (39) apparently can
account for simultaneous transport of water vapor and oxygen
through hydrophilic films. Variables related to the model such as
in-film diffusivities and concentration of penetrants, can be
determined by, for instance, well-defined water-vapor permeation
tests and easily installed routine computer programs.
Unfortunately, no experimental verification for the model was
provided. Judging from the advantages deduced from this model,
however, it is worthwhile in pursuing future research by applying
this modeling approach to other cases.

Generalized Model of Permeation with Moving Boundaries

Figgi (71) and Vom Bruck et al. (72) illustrated a schematic
model, shown in Figure 4, for a system of polymer and food having
an intermediate layer of swollen polymer. This model is initially
used to observe the migration of a penetrant, e.g. plastics
component, from the polymer phase to the fluid phase of food. The
mathematical description of the diffusion processes of the
penetrant in the homogeneous multiphase system with moving phase
boundaries has been developed by Rudolph (73,74). In fact, the
model is so generalized that it can also be used to describe the
oxygen transport through a polymeric film which swells due to
existence of water vapor.

In the development of the model, as shown in Figure 5, the
fluid phase F (a gas or liquid) containing a penetrant g occupies
a semi-infinite space from 0 to $+\infty$, while the polymer phase P
occupies the remaining region, 0 to $-\infty$. The interface between the
two phases is located at location $x = 0$.

During permeation, the polymer phase diffuses into the fluid
phase forming a swelling in the homogeneous region of mixed P and
F phases. Theoretically, two swelling fronts occur. One front
moves into the polymer phase, while the other moves into the fluid
phase, thereby changing the original volumes. Both fronts are
initially located at $x = 0$ and each moves with a velocity
inversely proportional to the square root of time. For
convenience, the locations of the swelling front for polymer
phase, x_p, is assumed as $x_p < 0$ and for fluid phase, $x_F > 0$.
Meanwhile, with the diffusion of polymer phase, penetrant g
diffuses out of fluid phase ($x_F < x < \infty$), into the swelling region
P+F ($x_p < x < x_F$) and eventually into polymer phase ($-\infty < x < x_p$).

Subscripts denoting the diffusing component (g, P or F) and
superscripts denoting the medium in which diffusion occurs (P, F,
or P+F) are used. For instance, D_g^{P+F} refers to the diffusivity of
penetrant g in the swelling region P+F.

Based on the concept of "two swelling fronts", Figure 4
illustrates a special case of a single swelling front by putting

$x_p = 0$. In other words, the volume of the mixture of P+F phases is assumed to be equal to that which was originally occupied by the polymer alone. The model thus defined belongs to the general class of diffusion problems with moving boundaries (16,5). The governing diffusion equations for the three regions mentioned above are:

$$\frac{\delta c_g^P}{\delta t} = D_g^P \frac{\delta^2 c_g^P}{\delta x^2} \qquad \text{for } -\infty < x < x_p(t) \qquad (35)$$

$$\frac{\delta c_g^{P+F}}{\delta t} = \frac{\delta}{\delta t} [D_g^{P+F}(c_P^{P+F}) \frac{\delta c_g^{P+F}}{\delta x}] \quad \text{for } -x_p(t) < x < x_F(t) \qquad (36)$$

$$\frac{\delta c_g^F}{\delta t} = D_g^F \frac{\delta^2 c_g^F}{\delta x^2} \qquad \text{for } x_F(t) < x < \infty \qquad (37)$$

The boundary and initial conditions are:

$$k_1 c_g^P = c_g^{P+F} \qquad \text{at } x = x_p \qquad (38)$$

$$c_g^{P+F} = k_2 c_g^F \qquad \text{at } x = x_F \qquad (39)$$

$$c_g^P = 0 \qquad \text{for } t = 0 \qquad (40)$$

$$c_g^F = C_g \qquad \text{for } t = 0 \qquad (41)$$

where k_1 and k_2 are the respective partition coefficient of penetrant g of (P+F)/P and (P+F)/F interfaces. C_g is the initial penetrant concentration in the fluid phase.

The assumption made for Equations 32-38 are:
- Diffusion is isothermal and one-dimensional,
- D_g^P, D_g^F, and D_P^F are constant (due to low penetrant concentration existing in both polymer and fluid phases),
- D_g^{P+F} in the swelling region depends on c_F^{P+F},

and
- The Nerst Partition Law governs the discontinuity of the penetrant concentration at the interfaces x_p and x_F.

Detailed discussion of the assumptions can be found elsewhere (74). Essentially, two diffusion fluxes are involved in the model; namely, (i) Fickian diffusion fluxes through the interface between different phases, and (ii) fluxes caused by the movement of the interfaces.

The governing equations are then coupled with the mass balance of penetrant g in the three regions:

$$M_g^P(t) + M_g^{P+F}(t) + M_g^F(t) = M_g^F(t = 0) \qquad (42)$$

and through the Boltzmann variable ($\underline{16}$)

$$\eta = x/2(t^{0.5}) \qquad (43)$$

and related transformation procedures to obtain general solutions ($\underline{71}$), which can be written in general form as ($\underline{69}$)

$$M_g^R(t) = \alpha c_g^P(t^{0.5}) \qquad (44)$$

where

$\alpha \quad = f[D_g^F, D_g^P, k_1, k_2, \eta_x, f^*(\eta_x, D_g^{P+F}, T)]$

$\eta_x \quad = f(k_F, c_P^P, c_P^F)$ (Boltzmann transformation)

$k_F \quad =$ Nernst partition coefficient for fluid phase

$T \quad =$ Temperature

$M_g^R(t) \quad =$ Amount of penetrant g in different regions R as a function of time t.

Apparently, Equation 41 indicates that $M_g^P(t)$ is proportional to the initial concentration of penetrant-rich fluid phase and to the square root of time. The concentration of the fluid phase in the polymer phase and of the penetrant g in all three regions of the system are dependent on the relation between location x and contact time t. Therefore, the time-dependent concentrations of fluid phase and penetrant g in the polymer phase can be plotted as a function (i.e. abscissa) of Boltzmann variable, η, defined in Equation 40.

In the case of constant D_g^{P+F}, the general solution can be easily modified to lead to the required equations. As pointed out by Rudolph ($\underline{73}$), by putting $D_g^{P+F} = D_g^P$ and $k_2 = 1$, $\eta_p = \eta_F = 0$, i.e. no diffusion of fluid phase into polymer phase, the general solution can be converted into the identical equation of so-called square root relationship for infinite and semi-infinite media given by ($\underline{16}$) for diffusion without swelling. If D_g^{P+F} is not a constant, $D_g^{P+F}(\eta)$ for the swelling region can then be mathematically described in four different cases as indicated in curves a through d in Figure 5. Curve a represents the case that the value of $D_g^{P+F}(\eta)$ is different from that of $D_g^{P+F}(\eta)$ and $D_g^F(\eta)$

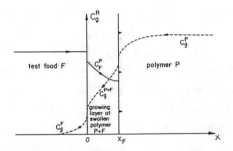

Figure 4. Model of the system polymer/test food. (Reproduced with permission from Ref. 72. Copyright 1986 Elsevier Applied Science.)

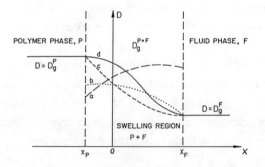

Figure 5. Various choice for the function $D_g^{P+F}(h)$ (schematic). (Reproduced with permission from Ref. 74. Copyright 1980 John Wiley & Sons.)

at $\eta = \eta_p$ and g_F, respectively. Curve b describes the condition that $D_g^{P+F}(\eta)$ equals to $D_g^P(\eta)$ at $\eta = \eta_p$, but not equals to $D_g^F(\eta)$ at $\eta = \eta_F$. Both curves c and d portray the condition that $D_g^{P+F}(\eta)$ equals to $D_g^P(\eta)$ and $D_g^F(\eta)$ at $\eta = \eta_p$ and η_F, respectively; however, the first derivative of $D_g^{P+F}(\eta)$ to η, $dD_g^{P+F}(\eta)/d\eta$, at $x = x_p$ and x_F are different. In curve c, $dD_g^{P+F}(\eta)/d\eta \neq 0$ and the curve is concave or convex, while in curve d, $dD_g^{P+F}(\eta)/d\eta = 0$ and the curve is sigmoidal. Therefore, depending upon given conditions, change in diffusivity of a penetrant in the swelling region can therefore be mathematically described.

If D_g^{P+F} is fixed by previously determined parameters such as D_g^P, D_g^F, k_1, k_2, η_p, and η_F, some special cases of the general solutions can be obtained. For instance, in the case of $D_g^F = \infty$ along with other with parameters, the fluid phase is treated as a well "stirred" medium. The polymer phase swells by the influence of the well-stirred fluid phase, while the penetrant g diffuses into the polymer phase, as shown in Figure 6. Noticeably, the jump in concentration of the penetrant g as a function of η is dependent on η_p or η_F, k_1 or k_2, and the respective square root of D_g^{P+F} and D_g^P.

In the cases of $D_g^P = \infty$, i.e. treating the polymer phase as a well stirred medium, only one swelling front at $\eta = \eta_F$ is considered, as shown in Figure 7. In the case of $D_g^P = 0$, then the diffusion of the penetrant g occurs only into that part of the polymer phase which is swollen, i.e. the swelling region, as shown in Figure 8.

Detailed solution forms of the above-mentioned cases along with other special ones such as the cases of $D_g^P = 0$, $D_g^{P+F} = \infty$, $D_g^P = \infty$ and $D_g^F = \infty$ were listed in (74). With properly selected parameters, these solutions can provide a unique model to describe the transient mass transfer of a penetrant inside the swelling region. Figgi et al. (75,76) reported the application of the above model to the results of diffusion experiments for the system of plastic packaging (polymer + additive) contents.

Since the general solutions of the model involves many integration constants, the calculation procedures are thus

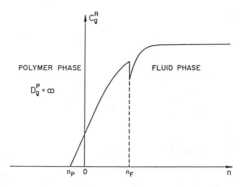

Figure 6.   Concentration of C_g^R of C_g in three regions. (R=P, P+F, F) vs. h with $D_g^F = \infty$ (schematic).   (Reproduced with permission from Ref. 74. Copyright 1980 John Wiley & Sons.)

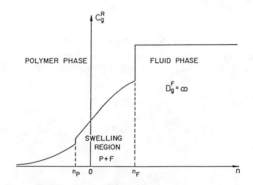

Figure 7.   Concentration of C_g^R of C_g in three regions (R=P, P+F, F) vs. h with $D_g^P = \infty$ (schematic).   (Reproduced with permission from Ref. 74. Copyright 1980 John Wiley & Sons.)

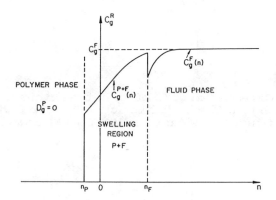

Figure 8.   Concentration C_g^R of C_g in three regions (R=P, P+F, F) vs. h with $D_g^P = 0$ (schematic).   (Reproduced with permission from Ref. 74. Copyright 1980 John Wiley & Sons.)

difficult and time-consuming. Also, some major constants such as partition coefficients of a penetrant in polymer and fluid phases are independent of the model and needed to be determined before using the general solutions of the model.

Concluding Remarks

Modeling of oxygen and water vapor permeation through polymeric film needs to consider whether the film is hydrophobic or hydrophilic. Based on the extent of the interaction between water vapor and the polymeric film, several modeling approaches were described. The governing diffusion equations used for these modeling approaches are Fickian.

The model developed by Peppas and Khanna (46) is a good model to account for both the packaged food and the polymeric film which are sensitive to vapor sorption. They also developed some generalized curves to predict the water activity and, therefore, the shelf-life of dehydrated food packaged in different films.

The model developed by Rudolph (74) is obviously the most generalized and comprehensive one. Although the model can predict the mass transfer of a penetrant with good estimation of the diffusivity of penetrant as function of swelling front movement, the general solution of the model, however, involves many variables and is time-consuming. Further, some variables such as the diffusivities of a given penetrant in the polymer and fluid phase, the partition coefficients at interfaces between different phases need to be individually determined before being applied to the model.

Meanwhile, the above-mentioned modeling approaches only consider the permeation of single penetrant and do not include the case of simultaneous oxygen and water vapor transport through a film except the model proposed by Schwartzberg (39). Although Schwartzberg's model is based on simplified assumptions related to surface adsorption and desorption of water, which may not be easily justified, the experimental set-up to determine the major variables of the model, such as the mass transfer coefficient of water vapor flux, the water vapor pressure at the evaporation interface of the polymeric film, is, however, well established (68). Also, the computer-aided calculation procedure and library software used are easy to install.

Although the above-mentioned modeling approaches have provided some satisfactory representation of oxygen and water vapor transport, it must be reminded that each approach is often based on rather drastic assumptions and the knowledge of the nature of the complex solubility-diffusion system is still incomplete. With continued effort further knowledge of all factors influencing the protective properties of a related film related to food quality will be gained and this may lead to the development of new packaging materials and a new modeling approach.

Literature Cited

1. Rizvi, S. S. H. CRC Crit. Rev. Food Sci. Nutri. 1981, 13, 111-33.
2. Mannheim, C; Passy, N. In Properties of Water in Foods in Relation to Quality and Stability; Simatos, D.; Multon, J. L., Ed.; Martinus Nijohoff: Dordrecht, The Netherlands, 1985; p 375.
3. Kester, J. J.; Fennema, O. R. Food Technol. 1986, 40 (12), 47.
4. Barrer, R. M. Diffusion In and Through Solids; 2nd. Ed.; Cambridge University: Cambridge, 1951.
5. Jost, W. Diffusion; Academic: New York, 1960.
6. Rogers, C. E. In Engineering Design for Plastics; Baer, E., Ed.; Reinhold: New York, 1964; Chap 9.
7. Rogers, C.E. In Physics and Chemistry of the Organic Solid State; Fox, D.; Labes, M. M.; Weissberger, A., Ed.; Interscience: New York, 1965; Vol. II, Chap 6.
8. Meares, P. Structure and Bulk Properties; Van Nostrand: London, 1965; Chap 12.
9. Crank, J.; Park, G. S. Diffusion in Polymers; Academic: New York, 1968.
10. Michaels, A. S.; Bixler, H. J. In Progress in Separation and Purification; Perry, E. S., Ed.; Interscience: New York, 1968; p 143-86.
11. Bixler, H. J.; Sweeting, O. J. In Science and Technology of Polymer Films; Sweeting, O. J., Ed.; Wiely-Interscience:New York, 1971 Chap 1.
12. Rogers, C. E.; Fels, M.; Li, N. N. In Recent Development in Separation Science;Li, N. N., Ed.; Chemical Rubber: Cleveland, 1972; Vol. II, p 197-255.
13. Stannett, V.; Hopfenberg, H. B.; Petropoulos, J. H. In MTP International Review of Macromolcular Science; Bawn, C. E. H., Ed.; Butterworth: London, 1972; p 329.
14. Hopfenberg, H. B. Permeability of Plastic Film and Coatings; Plenum: New York.
15. McGregor, R. Diffusion and Sorption in Fibers and Films; Academic: New York, 1974; Vol. 1.
16. Crank, J. The Mathematics of Diffusion; 2nd Ed.; Clarendon: Oxford, 1975.
17. Hwang, S. T.; Kammermeyer, K. Membranes in Separations; Wiley-Interscience: New York, 1975.
18. Meares, P. Membrane Separation Processes; Elsevier Scientific: Amsterdam, 1976.
19. Felder, R. M.; Huvard, G. S. Methods of Experimental Physics; Academic: New York, 1980; Part 16C, p 315-77.
20. Stern, S. A.; Frisch, H. L. Ann. Reviews of Materials Sci. 1981, 11, 223-50.
21. Rogers, C. E. In Polymer Permeability; Comyn, J., Ed., Elsevier: New York, 1985, p 11.
22. Stannett, V.; Yasuda, H. In Testing of Polymers; Smitz, J. V., Ed.; Interscience: New York, 1965.
23. Daynes, H. A. Proc. Roy. Soc. (London). 1920, A97, 273.

24. Barrer, R. M. Trans. Faraday Soc. 1939, 35, 628.
25. Michaels, A. S.; Vieth, W. R.; Barrie, J. A. J. Appl. Phys. 1963, 34, 1.
26. Machin, D.; Rogers, C. E. Crit. Rev. Macromol. Sci. 1960, 1, 245.
27. Pollack, H. D.; Frisch, H. L. J. Phys. Chem. 1959, 63, 1022.
28. Ziegel, K. D.; Frensdorff, H. K.; Blair, D. E. J. Polym. Sci. 1969, Part A-2, 7, 809.
29. Felder, R. M. J. Membr. Sci. 1978, 3, 15.
30. Jost, W. Diffusion in Solids, Liquids, and Gases; Academic: New York, 1952; p 30-45.
31. Rogers, C. E.; Stannett, V.; Szwarc, M. J. Polym. Sci. 1960, 45, 61.
32. Yasuda, H.; Rosengren, K. J. J. Appl. Polym. Sci. 1970, 14, 2839.
33. Pasternak, R. A.; Schimscheimer, J. F.; Heller, J. J. Polym. Poly. Sci. 1970, Part A-2, 8, 467.
34. Felder, R. M.; Spence, R. D.; Ferrell, J. K. J. Appl. Polym. Sci. 1975, 19, 3193.
35. Felder, R. M.; Ma, C-C.; Ferrell, J. K. AIChEJ. 1976, 22,724.
36. DeGroat, S. R.; Mazur, P. Non-equilibrium Thermodynamics. North-Holland: Amsterdam, 1962.
37. Katchalsky, A.; Curran, P. F. Non-equilibrium Thermodynamics in Biophysics. Harvard University: Cambridge, 1967.
38. Cussler, E. L. Multicomponent Diffusion. Elsevier: New York, 1976.
39. Schwartzberg, H. G. In Food Packaging and Preservation: Theory and Preservation; Elsevier: New York, 1986. p.115.
40. Barrer, R. M.; Skirrow, G. J. Polymer Sci. 1948, 3, 549.
41. Alter, H. J. Polymer Sci. 1962, 57, 925.
42. Michaels, A. S.; Bixler, H. J. J. Polym. Sci. 1961, 50 413.
43. Jasse, B. In Food Packaging and Preservation: Theory and Practice; Elsevier: New York, 1986, p.316-18.
44. Michaels, A. S.; Vieth, N. R.; Bixler, H. J. J. Appl. Polymer. Sci. 1964, 3, 2735.
45. Yasuda, H.; Peterline, A. J. Appl. Polym. Sci. 1974, 18, 531.
46. Peppas, N. A.; Khanna, R. Polym. Eng. Sci. 1980, 20, 1147.
47. Lightfoot, E. Transport Phenomena and Living Systems; Wiley: New York, 1974.
48. Bird, R. B.; Steward, W. E.; Lightfoot, E. N. In Transport Phenomena; John Wiley & Sons: New York, 1960, 569-72.
49. Brandrup, J.; Immergut, E. H. Polymer Handbook; Wiley: New York, 1975.
50. Guggenheim, E. A. Application of Statistical Mechanics; Clarenden: Oxford, 1966.
51. Anderson, R. B. J. Am. Chem. Soc. 1946, 68, 686.
52. de Boer, J. H. The Dynamical Character of Adsorption; Clarenden: Oxford, 1953.
53. Van den Berg, C. In Engineering and Food, vol.1:Engineering Science in the Food Industry; McKenna, B. M., Ed.; Elsevier: New York, 1984; p 311-21.

54. Labuza, T. P.; Kaanane, A.; Chen, J. Y. J. Food Sci. 1985, 50, 385.
55. Rotstein, E.; Cornish, A. R. H. AIChE J. 1978, 24, 956.
56. Karel, M.; Proctor, B. E.; Wiseman, G. Food Technol. 1959, 13, 69.
57. Quast, D.; Karel, M.; Rand, W. M. J. Food Sci. 1972, 37, 673.
58. Quast, D; Karel, M. J. Food Technol. 1971, 6, 95.
59. Mizrahi, S.; Karel, M. J. Food Sci. 1978, 43, 750.
60. Mizrahi, S.; Karel, M. J. Food Technol. 1977, 42, 95.
61. Khanna, R.; Peppas, N. A. AIChE Symposium Series. 1982 78(218), 185.
62. Simon, I. B. M.S. Thesis. Massachusetts Institute of Technology, Cambridge, 1969.
63. Salwin, H.; Slawson, V. Food Technol. 1959, 13, 715.
64. Yasuda, H.; Lamaze, C. E. J. Maacromol. Sci. - Phys. 1971, 85, 111.
65. Kokes, R. J.; Long, F. A. J. Am. Chem. Soc. 1953, 75, 1255.
66. Hauser, P. M.; McLaren, A. D. Ind. Eng. Chem. 1948 40, 112.
67. Karel, M.; Proctor, B. E.; Wiseman, G. Food Technol. 1959, 13 (1), 69.
68. Annual Book of ASTM Standards; ASTM: Philadelphia, 1983; Sec. 15.09, vol. 15.
69. Robinson, B. Statistical Package for the Social Sciences; Vogelback Computing Center, Northwestern: Chicago, 1983.
70. Chao, R. Ph.D. Thesis, University of Massachusetts, Amherst, 1984.
71. Figgi, K. Progress in Polym. Sci. 1980, 6, 187.
72. Vom Bruck, C. G.; Bieber, W. D.; Figgi, K. In Food Packaging and Preservation: Theory and Practice; Mathlouthi, M., Ed.; Elseiver: New York, 1986; p39.
73. Rudolph, F. B. J. of Polym. Sci., Polym. Phys. Ed. 1979, 17, 1709.
74. Rudolph, F. B. J. of Polym. Sci., Polym Phys. Ed. 1980, 18, 2323.
75. Figgi, K.; Rudolph, F. B. Angew. Makromol. Chem. 1979, 78, 157.
76. Figgi, K.; Klahn, J. Angew. Makromol. Chem. 1982, 107, 117.

RECEIVED December 11, 1987

Chapter 19

Moisture Transfer and Shelf Life of Packaged Foods

P. S. Taoukis, A. El Meskine, and T. P. Labuza

Department of Food Science and Nutrition, University of Minnesota, St. Paul, MN 55108

Control of moisture exchange with the environment, accomplished through packaging, is crucial for moisture sensitive foods. Moisture transport prediction models for packaged foods, based on the linear and GAB isotherms, are presented. The models can be used for the selection of the appropriate food-package systems, for the estimation of the effects of changes in the package or environment parameters and for the prediction of the shelf life of the product under variable conditions. A dimensionless number, L, that can serve as a simple quantitative criterion for the applicability of the packaging models, is introduced. L relates the resistances to moisture transport through the film barrier and inside the food itself and is used to test the models' assumption that the film resistance is the dominant one.

Foods are complex biologically and chemically active systems that require strict control of their manufacturing, distribution and storage conditions in order to maintain their safety and their sensory and nutritive values. The time period, from processing, during which a food stays within acceptable limits of quality is defined as shelf life. To ensure a high quality product for at least the extent of its targeted shelf life, environmental conditions such as temperature, moisture, gas composition and light and changes thereof have to be accounted for and, if possible, controlled. Packaging is one of the means employed to accomplish that, especially with regard to moisture, gas composition and light. The primary function of a food package, besides serving as the containing unit, is to keep the food in a controlled microenvironment. The package itself becomes part of the food's environment, and the food-package interactions have to be considered. The paper focuses on the role of packaging in controlling the water content of the food and its effects on shelf life.

Water In Foods

Water is of major importance in food preservation. Reduction of available water is the basis of some of the earliest preservation techniques. In the last 30 years the physico-chemical and biological principles governing the mechanisms of water-food interactions have been systematically investigated (1). The single most important physico-chemical property in food systems is the water activity, a_w, of the food defined from the thermodynamic equilibrium state (2). Under normal pressure conditions, a_w is practically equal to the ratio of the water vapor pressure of the food, p, at equilibrium divided by the vapor pressure of the pure water, p_o, at the same temperature and is related to the equilibrium relative humidity (% ERH):

$$a_w = p/p_o = \%ERH/100 \qquad (1)$$

At a constant temperature a unique relation exists between moisture content and a_w of a specific food, depending on its method of preparation (i.e. adsorption versus desorption). This relation is depicted by the moisture sorption isotherm of the food. In Figure 1 a typical food moisture isotherm is shown. The range of a_w for different types of food is also presented. A comprehensive treatment of moisture sorption is given by Labuza (1984) (3).

Moisture Content and Food Stability

Water activity describes the degree of boundness of the water contained in the food (4) and its availability to act as a solvent and participate in chemical or biochemical reactions. Critical levels of a_w can be recognized above which undesirable deterioration of food occurs, from a safety or quality point of view. From a safety standpoint, we are basically concerned with microbial growth. The ability of a microorganism to grow in a given environment depends on the complex interactions of a number of factors: water activity, temperature, pH, oxidation-reduction potential, preservatives and competitive microflora (5, 6). For set values of the other factors, a minimum a_w for growth can be defined for a given microbial species. The most tolerant pathogenic bacterium is Staphylococcus aureus, which can grow down to an a_w of 0.84-0.85. This is often used as the critical level of pathogenicity in foods. This limit is pertinent to intermediate moisture foods (IMF) that are manufactured at that range of a_w (7). Xerophilic molds and osmophilic yeasts can grow down to 0.6-0.7 a_w and can become of importance in dry foods. Beuchat (1981) (8) gives minimum a_w values for a number of commonly encountered microorganisms of public health signifi- cance. Richard-Molard et al. (1985) (9) reviewed the effect of a_w on molds and give mimima for growth and mycotoxin production, and Tilbury (1976) (10) pre- sented a comprehensive treatment of a_w tolerant yeasts. The Food and Drug Administration (FDA) in the recently revised Current Good Manufacturing Practice regulations (CGMP) under 21 CFR 110.3(n) defined "safe moisture level" as a level of moisture low enough to prevent the growth of undesirable microorganisms in the finished product under the intended conditions of manufacturing, storage and distribution. The maximum safe moisture level is based on the a_w of the food. An a_w is considered safe if adequate data are available that demonstrate the food at or below the given a_w will not support the growth of undesirable microorganisms. In 21 CFR 110.80(b)(14) the FDA requires that intermediate moisture and dry foods that rely on a_w to prevent growth of undesirable microorganisms should be processed to and maintained at a safe moisture level. One means to do this is 21 CFR 110.80 (b)(14)(iii) "Protecting finished food from moisture pickup, by means of a moisture barrier or by other means...."

Textural quality is also greatly affected by moisture content and water activity. Dry, crisp foods like potato chips, popcorn, crackers and cornflakes

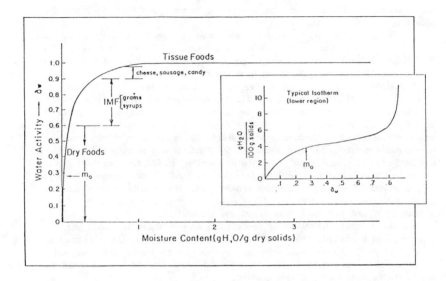

Figure 1. Example of a typical food isotherm. Also the water activity ranges of different food categories is shown.

lose crispness and become texturally unacceptable upon gaining moisture usually
in the a_w range of 0.35 to 0.5 (12). Intermediate moisture foods like dried
fruits, pet foods, bakery goods and some confectionery items upon losing moisture
below the range of 0.5 to 0.7 a_w, become unacceptably hard (13). Unpopped popcorn below 0.3 to 0.4 a_w will result in a significantly reduced popping volume
(14). Pasta, especially larger cuts like lasagna noodles, becomes prone to
cracking (checking) below 0.4 (15). Another serious problem is encountered in
the case of products containing amorphous sugars, such as many spray-dried foods
that contain soluble carbohydrates. When the a_w increases above 0.35 to 0.4,
amorphous sugars cake and recrystallize releasing water and thus significantly
affect texture and quality (16).

Another critical a_w limit can be set based on economic and regulatory considerations. Foods losing moisture beyond a "reasonable" level (not well defined) become illegal from a net weight point of view (17). The critical a_w will
be variable in this case, depending upon the food isotherm.

Besides the specific critical a_w limits, water activity has a pronounced
effect on chemical reactions. The ability of water to act as a solvent, reaction medium and as a reactant itself increases with increasing a_w. As a result,
many deteriorative reactions increase exponentially in rate with a_w above the
value corresponding to the monolayer moisture, the value at which most reactions
have a minimum rate (18). The above can be schematically represented in a global food stability map (Figure 2). The critical a_w limits for microbial growth
and the relative rates of reactions important to food preservation such as lipid
oxidation and non-enzymatic browning can be seen. Most reactions have minimal
rates up to the monolayer value. Lipid oxidation shows the peculiarity of a
minimum at m_0 and increased rates below and above it.

From the above discussion on the significance of water in foods, the
importance of being able to maintain a food within certain a_w limits through use
of appropriate packaging is apparent and the ability to predict the change of
moisture and a_w throughout shelf-life under different storage conditions is a
key to maintaining packaged food quality.

Moisture Transport in Packaged Foods

In most packaged foods the major moisture transport mechanisms are water vapor
diffusion through the packaging barrier and within the food (Figure 3). Usually
the transport through the film barrier controls the phenomenon. The moisture
transport through a thin packaging film is described by a pseudo-steady state
equation based on Fick's and Henry's laws:

$$W_S \frac{dm}{dt} = \frac{k}{x} A(p_e - p) \tag{2}$$

where dm/dt = the rate of moisture per unit dry weight transferred per day
          k/x = the film permeance to moisture (in g/day m^2 mmHg)
          A = the effective area of diffusion
          W_S = the food's total dry weight in the package
          p, p_e = water vapor partial pressure in the package and in the
                  environment, respectively

This equation is valid for a thin, non-porous, hydrophobic film, with low water
solubility, non-swelling and with constant permeance over time. The package is
assumed sealed without a total pressure difference from the environment (19).

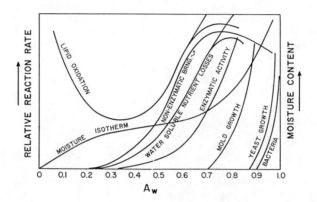

Figure 2. Global food stability map.

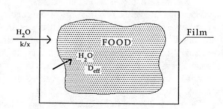

Figure 3. Moisture transfer in a food-package system.

Permeance measurements.   The standard method of film permeance measurement
(ASTM-E-96) is a method in which a metal cup is filled with desiccant, the cup
is covered and sealed with the film and the weight gain under constant conditions of 100°F (37.7°C) and 90% RH is recorded. The vapor pressure gradient,
Δp, is constant and integration of Equation 2 gives a linear relation of
moisture vs. time, from the slope of which the film permeance can be calculated.
Caution should be drawn to the fact that often in the available literature,
instead of the permeance, values of water vapor transmission rate (WVTR) are
given where: WVTR = slope x A = k/x Δp. A number of other methods for permeance
measurements can also be used (20). The most prominent uses a device that
measures the vapor pressure change with time with a moisture vapor sensor. It
gives the permeance value directly as a readout (Mocon Instruments). The typical range of moisture permeances of different packaging materials is shown in
Table I.

Table I. Film Permeances at 35°C

$$\frac{k}{x} = \frac{Kg\ water}{m^2\ hr\ mm\ Hg}\ x10^{-7}$$

Paperboard	3333
Polypropylene	137
Cellophane/polyethylene	102
1 mil polyethylene	86
Polyethylene-terephthalate	50
Polyester	33
PET/PE	19
Polyester/foil/PE	1

100% RH driving force

Permeance Parameters. A number of factors affect the permeance of packaging
materials. These factors were reviewed by Salame and Steingiser (1977) (21) and
Pascat (1986) (22) and will be treated in a following chapter. To summarize,
with regard to moisture, permeance depends on the nature of the polymeric film.
Varying functional groups in the repeating unit give different k/x, depending on
their polarity and stereochemical properties. An increase in crystallinity,
molecular orientation, density and molecular weight results in decrease of permeance. An increase in double bonds and polarity results in increased permeance. Additives, plasticisers, inert fillers, reticulation and irradiation
also affect k/x. Permeance is also affected by thickness. Fick's Law considers
permeance to be inversely proportional to the thickness, x, but in general it is
proportional to x^{-n}, where n = 0.8 to 1.2 (23). Most importantly, permeance
depends on temperature, T, and external relative humidity, %RH. In general, it
follows the classical Arrhenius relationship with the activation energy, E_A,
being a function of RH where E_A ranges from 5 to 20 kcal/mol (23). As a rule of
thumb, permeance increases by 30 to 50% for every 5°C (21). In Figure 4, an
Arrhenius plot of log(k/x) versus 1/T for a packaging film showing the effect of
external %RH is presented. The temperature dependency (E_A) changes when the
material goes below its glass transition temperature (Tg), thus the permeance
data at temperatures above Tg should not be extrapolated below that temperature
(24, 25).
      Permeance values and related information are available in a number of
literature sources compiled in tables (26), or nomagraphs (27) or in the form of

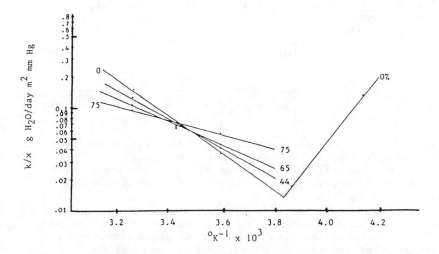

Figure 4. Arrhenius plot of ℓn k/x versus reciprocal absolute temperature for polyethylene terephthalate/polyethylene laminate illustrating effect of external %RH on film permeance.

prediction equations (21, 28). Manufacturers of packaging materials also provide relevant information. Nevertheless, the large number of parameters that
affect permeance including stress and stretching during the actual package formation may result in significant deviations from the quoted permeance values.
The effect of pinholes has also been considered and studied (29). It is thus
often advisable to measure permeance of the formed package using the foremen-
tioned gravimetric method with desiccant being put into the package, sealed and
stored at relevant temperature and humidity.

Packaging Equations

If we assume that the vapor pressure in the package is in equilibrium with the
food, the moisture transport equation (Equation 2) can be integrated (for
constant external temperature and relative humidity) to give the following
packaging equations:

$$\int_{m_i}^{m} F(m) \ dm = \frac{k}{x} \frac{A}{W_s} \ t \tag{3}$$

or

$$\int_{a_{wi}}^{a_w} G(a_w) \ da_w = \frac{k}{x} \frac{A}{W_s} \ t \tag{4}$$

where $F(m)$ and $G(a_w)$ are functions that depend on the form of sorption isotherm
equation used. A number of different isotherm equations have been used for
derivation of $F(m)$ or $G(a_w)$ and use in Equations 3 or 4, by various researchers
in the area of packaging predictions. A compilation of some of these equations
is shown in Table II.

Linear Isotherm Solution. No single isotherm equation is univerally acceptable
and each equation is applicable at a certain a_w range and for certain categories
of foods. It is noted that only for the linear and Langmuir isotherm equations
have analytical solutions to the packaging model been presented. The other
solutions are based on numerical integration of Equations 3 or 4.

The linear approximation of the food isotherm, usually good between 0.2
to 0.6 a_w, was first used by Karel and Labuza (1969) (32) as seen in Table II.
Using the linear isotherm, Equation 2 can be integrated to Equation 3, where
$F(m) = b/p_o(m_e-m)$, and a simple analytical solution can be given:

$$\ln \Gamma = \phi_{ext} \ t \tag{5}$$

with $\Gamma = \dfrac{m_e - m_i}{m_e - m}$ (6a) $\phi_{ext} = \dfrac{k}{x} \dfrac{A}{W_s} \dfrac{P_o}{b}$ (6b)

In this case, m_i is the initial moisture content of the food (dry basis) and m_e
is the equilibrium moisture corresponding to the external RH from the linear
isotherm. Γ is the unaccomplished moisture fraction and ϕ_{ext} the overall permeance of the system. Equation 6 shows that ϕ_{ext} depends on the film's properties and thickness, the environment temperature, the slope of the isotherm and
the A/W_s ratio. The linear model has been extensively used in our laboratory
and by other researchers (34, 35) and has been found to be in very good
agreement with experimental results, especially in the range where the linear
isotherm closely follows the actual isotherm.

Table II. Isotherms used in packaging predictions

Isotherm equation	Form	Reference
Oswin	$m = b\left[\dfrac{a_w}{1-a_w}\right]^c$	30
Graphic fit	————	31
Linear	$m = ba_w + c$	32, 33, 34, 35, 19
Two parameter	$a_w = \dfrac{b+m}{c+m}$	33
Kuhn	$m = \dfrac{b}{\ln a_w} + c$	33
BET	$m = \dfrac{m_0 \, c \, a_w}{(1-a_w)(1-a_w + ca_w)}$	36, 19
Langmuir	$\dfrac{a_w}{m} = c + \dfrac{a_w}{m_0}$	19
Freudlich	$m = b \, a_w$	19
Hasley	$a_w = \exp[-b/m^c]$	19, 37

GAB Isotherm Solution.   A moisture sorption isotherm equation that has lately
emerged as applicable to most foods is the Guggenheim-Andersen-DeBoehr or GAB
isotherm equation ($\underline{38}$):

$$m = \frac{m_0 \ C \ k \ a_w}{(1-ka_w) \ (1-ka_w+Cka_w)} \tag{7}$$

It is a three-parameter equation based on the BET sorption theory and developed
independently on principles of statistical mechanics and kinetics. The parameters of the equation have physical significance, m_0 being a monolayer moisture
value and C and k relating to interaction energies between water and food and
between the multiple layers of water respectively. For k=1 the equation reduces
to the well-known BET sorption isotherm equation. The GAB equation applies very
successfully to a large number of foods in the range of 0 to 0.9 a_w. It has
been shown that the GAB equation fits food isotherms in that range as well or
better than other equations with four or more parameters ($\underline{39}$). Figure 5
illustrates an isotherm for sodium caseinate with the GAB equation fit shown.
The GAB equation can also describe water activity dependence on temperature
since m_0, C, k are exponential functions of inverse absolute tempearture (1/T)
($\underline{40}$). It has become the standard sorption isotherm equation used in Europe
(E.E.C. COST 90 project on water activity) and is being established in U.S.
laboratories.
        The packaging prediction equation, using the GAB equation, has been solved
by us and takes the following form:

$$\int_{a_i}^{a} \frac{1 + k^2(C-1)a^2}{(a_e-a)(1-ka)^2 \ (1-ka+Cka)^2} \ da = \frac{P_0}{m_0CK} \ \frac{k}{x} \ \frac{A}{W_s} \ t \tag{8}$$

where a is the water activity of the food at time t and a_i the initial water
activity. Equation 8 can be solved both numerically and analytically. Using
this equation, the predicted increase in moisture and water activity with time
for a packaged sodium caseinate sample is shown in Figure 6. It can be seen
that up to 0.5 a_w, where the isotherm is almost linear, predictions from the two
models almost coincide. Above that, through equilibrium, the GAB and the linear
models deviate because the linear isotherm cannot describe the actual isotherm
very well in this range. In general, in comparing the two packaging models it
can be shown that they are equally accurate for the linear portion of the
isotherm while the GAB model is more accurate closer to equilibrium and can be
used to predict equilibrium time. The linear model is simpler to use however.
Nevertheless, there is difficulty in defining the "best" linear fit to the
isotherm which is usually different for different initial and critical conditions. The GAB equation with the three parameters defines the whole isotherm.
Literature values can be easily used for a quick packaging estimation.
Tabulation of GAB isotherms for more than 160 foods has been published from our
laboratory ($\underline{41}$, $\underline{42}$).

Use of Packaging Models

The moisture transport prediction models can be used to determine the optimum
package system to keep the product within certain a_w limits for its shelf-life

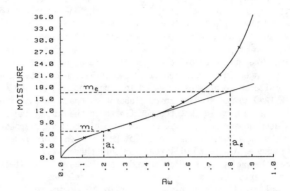

Figure 5. Sodium caseinate isotherm. Discrete points are experimental data (from Ref. 40). Continuous lines are the linear and the GAB equation fitted to the isotherm data.

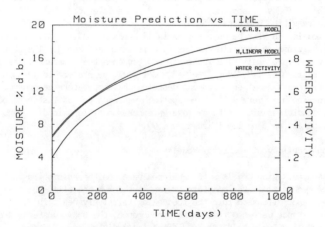

Figure 6. Moisture increase prediction for sodium caseinate of 0.2 initial a_w, packaged in 1 mil polyethylene and stored at 75% RH, 25°C.

period. Different overall permeances can be used to determine the best combination. In Figure 7, the a_w change with time for sodium caseinate in different packages is shown. It can be seen that if the objective is to keep the
product below 0.6 for one year the best film is polyethylene, with the paper
being inadequate and the PET/polyethylene overprotecting the system.

The models can also be used to predict the effect of change of parameters
like external RH, area of package to foods weight ratio (A/W_s) and temperature,
for the same film. The effect of different external relative humidities on
sodium caseinate packaged in polyethylene is shown in Figure 8. The effect of
change of package size on cornflakes packaged in a paperboard box with a
glassine liner (WVTR=0.1) was studied in our laboratory based on experimental
isotherm data. It was estimated that the cornflakes in the large size package
(A/W_s=0.3), will keep under the 0.4 critical water activity for crispness for
more than six months in a humid environment (75% RH), whereas cornflakes in the
individual size package (A/W_s=0.6) will keep for only three months.

Furthermore, the packaging prediction model can be used in shelf-life predictions of foods deteriorating due to water activity sensitive reactions like
non-enzymatic browning (43) or ascorbic acid loss in IMF or dry foods. For
example, the effect of a_w on non-enzymatic browning rates of dehydrated cabbage
was studied and modeled by Mizrahi and co-workers (1970) (44). The study showed
an increasing browning rate with an increase in a_w. Using this information and
the packaging models, one can predict the extent of browning with time for different packaging materials and storage conditions. Data were presented for two
different films (45). Experimental values were in very good agreement with predicted ones. Similar studies and predictions were published for ascorbic acid
loss (45, 36). The same approach can be used for accelerated shelf-life testing
of moisture sensitive products (45, 47).

The models can also be used for variable storage conditions if the effect
of these conditions on the film and the food isotherm are known. The storage
time is divided into small time intervals and for each the constant storage conditions model is used (48, 49). The GAB model would be preferable in such an
application since the GAB equation can describe the temperature dependence of
the sorption isotherm very well. The approach was used to model storage of wild
rice under conditions simulating average warehouse conditions in different U.S.
cities during a year's period (Figure 9) (50).

Test of Applicability of Packaging Models

The discussed packaging models were developed based on the assumption that the
controlling mechanism is the moisture transport through the film, the resistance
of moisture vapor diffusion into the food being relatively negligible. If the
internal resistance were controlling the phenomenon, the shown unsteady state
diffusion equation would hold (for one dimensional diffusion):

$$\frac{\partial m}{\partial t} = \frac{\beta_D}{\rho} \frac{\partial^2 P}{\partial x^2} = D_{eff} \frac{\partial^2 m}{\partial x^2} \qquad (9)$$

where β_D is the permeability of the food (kg/m hr mmHg), ρ is the density of the
food and D_{eff} is the effective diffusivity of moisture in the food (m^2/hr.).

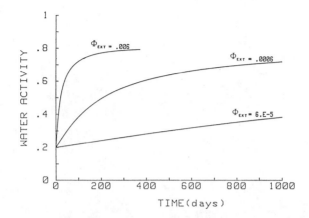

Figure 7. A_W increase prediction for sodium caseinate packaged in lined paper, polyethylene and PET/polyethylene (respectively from top), stored at 80% RH, 25°C.

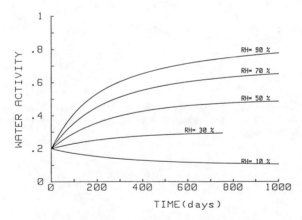

Figure 8. A_W change prediction for sodium caseinate packaged in polyethylene and stored at different ambient relative humidities (at 25°C).

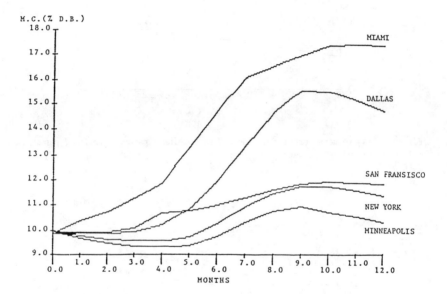

Figure 9. Predicted change in moisture content of packaged wild rice under the different storage conditions that would be encountered in five cities starting with production on January 1.

Using the linear isotherm and Equation 9, a relation between food permeability and effective diffusivity can be deduced:

$$D_{eff} = \frac{\beta_D}{\rho} \frac{P_o}{b} \tag{10}$$

Under the simplifying assumptions of negligible surface resistance, no change in D_{eff} with moisture content, slab geometry of 2 L_o total thickness and long times (such that $D_{eff} t/L_o^2 > 0.1$) an approximate analytical solution to Equation 9 (51 – 53) can be written as:

$$\ln \Gamma = \phi_{int} t + \ln \frac{\pi^2}{8} \tag{11}$$

where

$$\phi_{int} = \frac{D_{eff} \pi^2}{4 L_o^2} \tag{12}$$

Equation 11 can be used to estimate D_{eff} by a simple gravimetric experiment where the moisture increase of the slab shaped food stored under constant relative humidity conditions is measured with time. The slope of $\ln \Gamma$ vs time (Equation 12) is used to calculate D_{eff}. From it, if the food isotherm is known, the food permeability is calculated. Alternatively, β_D can be measured in a steady state permeability apparatus (54). This involves the very difficult task of shaping the food into a film and very few measurements of this type are reported in the literature.

One can assess the relative importance of the two mechanisms by comparing the values of the external permeance, ϕ_{ext} and the internal permeance ϕ_{int} (55). ϕ_{ext} and ϕ_{int} are equal to the rate of change of $\ln \Gamma$ in the cases where the transport through the film or the internal diffusion mechanism is controlling, respectively from Equations 5 and 11. If the rate of change of $\ln \Gamma$ due to one mechanism is much larger than the one due to the other mechanism, then the first mechanism can be neglected without significant error being introduced. To put that into quantitative terms, the ratio of the two permeances is considered. From Equations 6, 10 and 12, this ratio can be approximately written as:

$$\frac{\phi_{int}}{\phi_{ext}} \simeq 5 \frac{\beta_D/L_o}{k/x} \tag{13}$$

We can define the dimensionless number $L_\#$ as:

$$L_\# = \frac{\beta_D/L_o}{k/x} \tag{14}$$

This is similar to the Sherwood number used in mass transfer, which is a ratio of internal to external mass transfer. The L number can be used as a criterion for the applicability of the simple external mass transfer packaging models. If $\phi_{int} > 20 \phi_{ext}$ then the effect of the internal diffusion can be neglected. Based on this the following criteria can be used: If $L_\# > 4$, the film is the major resistance to moisture transfer and the packaging models apply well; if $0.2 < L_\# < 4$, the diffusion mechanism introduces error but the film barrier model can still be used if overestimation of the transported moisture is acceptable;

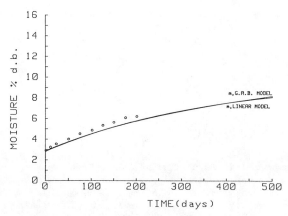

Figure 10. Moisture gain of cornflakes packaged in polystyrene. Points are experimental data and continuous lines are predictions using packaging models. Parameters: $k/x=3.334 \times 10^{-6}$ kg/m^2 hr mmHg, $D_{eff}=4.68 \times 10^{-6}$ m^2/hr., $\beta_D=2.44 \times 10^{-5}$ kg/m^2 hr mmHg., $L_o = 7$ cm.

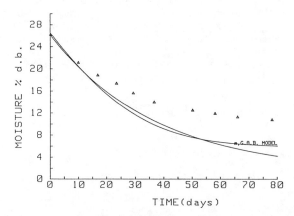

Figure 11. Moisture loss of dates packaged in cellophane. Points are experimental data and continuous lines are packaging model predictions. Parameters: $k/x=3.61 \times 10^{-4}$ kg/m^2 hr mmHg, $D_{eff}=4.2 \times 10^{-7}$ m^2/hr, $\beta_D=2.6 \times 10^{-5}$ kg/m^2 hr mmHg, L_o-7 cm.

if the L<0.2, the diffusion in the food mechanism is dominant and the use of the packaging model is not recommended. A more complex model involving the simultaneous numerical solution of the two transport equations (Equations 2 and 9) should be used.

Figures 10 and 11 illustrate the applicability of the L number criterion. In these figures, prediction values are compared to experimental data generated in our laboratory (56). Corn flakes packaged in polystyrene film (Clysar 125 EHC) with L_{ff}=10.4, stored at 35°C and 75% RH, showed a moisture increase that was very closely predicted by the packaging models (Figure 10). On the other hand, the models performed poorly in the case of dates stored in a cellophane (195 PSD) film. The L number of this second system is 0.10 and the packaging models overestimated the moisture lost by the dates (Figure 11). In Table III the L number is given for a variety of foods packaged in three materials covering the whole permeance range and for two package thicknesses (L_0=1 cm and 10 cm). The permeabilities of the foods were calculated from D_{eff} values measured in our laboratory (57). For the high barrier films, foil and polyethylene, the L number is in the applicability range of the packaging model in almost all cases. For low barrier materials like lined paperboard and cellophane the model will often not be accurate, especially for foods with a low effective diffusivity, and a more complex model may be needed.

Table III. Values of L Number for a Variety of Packaged Foods

PRODUCT	$\beta_D \times 10^{-5}$	L Number					
		$(k/x)^c = 1 \times 10^{-7}$		$(k/x) = 5 \times 10^{-6}$		$(k/x) = 1 \times 10^{-4}$	
		FOIL		POLYETHYLENE		PAPERBOARD	
		1 cm	10 cm	1 cm	10 cm	1 cm	10 cm
Flour	0.19[a]	1900	190	95	9.5	1.9	0.19
NFDM	0.09	900	90	45	4.5	0.9	0.09
Dried Vegetable	0.03	300	30	15	1.5	0.3	0.03
Peanut	0.02	200	20	10	1.0	0.2	0.02
Cookie	0.017	170	17	8.5	0.85	0.17	0.017
Raisin	0.002	20	2	2	0.1	0.02	0.002
Turkey[b]	4.5	45000	4500	2250	225	45	4.5

[a]Values calculated from D_{eff} data and an average isotherm slope b=0.25; in kg/m hr mmHg
[b]Experimental permeability value for freeze dried material; [c]values in kg/m² hr mmHg

Acknowledgments

This study was supported by University of Minnesota Agricultural Experiment Station Projects 18-72 and 18-78. Mr. El Meskine was supported by the US AID Morocco project with Hassan II University, Rabat. This is scientific series paper number 15,546 from the University of Minnesota Agricultural Experiment Station.

Literature Cited

1. Troller, J.A.; Christian J.H.B. Water Activity and Food; Academic Press: New 4York, 1978.
2. Van den Berg, C.; Bruin, S. in Water Activity: Influences in Food Quality; Rockland L., Stewart G. F., Eds.; Academic Press: New York, 1981 p. 1-64.

3. Labuza, T.P. Moisture Sorption: Practical Aspects of Isotherm Measurement and Use; American Association of Cereal Chemists: St. Paul, MN 1984.
4. Labuza, T.P. J. Food Proc. Preserv. 1977, 1, 167-190.
5. Leistner, L. and Rodel, W. Intermediate Moisture Foods; Davies R., Birch G.G., Parker J., Eds.; Applied Science: London, 1976 p. 120-37.
6. Troller, J.A. Food Technology 1980, 34(5), 76-80.
7. Taoukis, P.S.; Breene, W.M.; Labuza, T.P. in Advances in Cereal Science and Technology, Vol. VIII; Pomeranz Y., Ed.; American Association of Cereal Chemists: St. Paul, MN 1987.
8. Beuchat, L. Cereal Foods World. 1981, 26, 345.
9. Richard-Molard, D.; Lesage, L.; Cahagnier, B. in Properties of Water in Foods; Simatos D., Multon J.L. Eds.; Nijhoff Publishers: Dodrecht, Netherlands, 1985, p. 273-92.
10. Tilbury, R.H. In Intermediate Moisture Foods; Davies, R., Birch, G.G., Parker J., Eds.; Applied Science: London, 1976, p. 138-65.
11. 51 Federal Register 22458, 1986.
12. Katz, E.; Labuza, T.P. J. Food Sci. 1981, 46, 403.
13. Kochhar, S.P.; Rossell, J.B. J. Food Tech. 1982, 17, 661-668.
14. Brown, G.K. Smoke Signal 1968, 3, 68.
15. Winston, J.J. Macaroni, Noodles and Pasta Products; Winston Publishing Co.: New York, 1971.
16. Saltmarch, M.; Labuza, T.P. J. Food Science, 1980, 45, 1231.
17. Labuza, T.P. Food Technol., 1982, 36(4), 92-97.
18. Labuza, T.P. 1980. Food Technol. 1980, 34(4), 36-41.
19. Peppas, N.A.; Khanna, R. Polymer Eng. Sci., 1980, 20, 1147-1156.
20. Holland, R.V.; Santangelo, R.A. CSIRO Fd. Res.Q. 1984, 44, 20-22.
21. Salame, M.; Steingiser, S. Polym. Plast. Technol. Eng. 1977, 8(2), 155-75.
22. Pascat, B. In Food Packaging and Preservation: Theory and Practice; Mathlouthi, M., Ed..; Elsevier Applied Science: London, 1986, p. 7-24.
23. Labuza, T.P.; Contreras-Medellin, R. Cereal Foods World, 1981., 26(7), 335-343.
24. Frisch, H.L. Polymer Eng. Sci. 1980, 20, 2-13.
25. Weinhold, S. Proc. ACS Div. Polym. Mat. Sci. Eng., 1987, 56, 47.
26. Bakker, E. Encyclopedia of Packaging Technology; Wiley: New York, 1986.
27. Neitzert, W.A. Plastverarbeiter. 1979, 30(12), 773-447.
28. Lee, W.M. Polymer Eng. Sci., 1980, 20, 65-9.
29. Becker, K. Verpack. Rundsch. 1979, 12, 87.
30. Oswin, C.R. J. Soc. Chem. Ind. 1946, 65, 419.
31. Heiss, R. Modern Packaging 1958, 31(8), 119.
32. Karel, M.; Labuza, T.P. Optimization of protective packaging of space foods; U.S. Air Force contract F-43-609-68-C-0015. Aerospace Med. School, San Antonio, TX, 1969.
33. Labuza, T.P.; Mizrahi, S.; Karel, M. Trans. ASAE 1972, 15, 150.
34. Veillard, M.; Bentejac, R.; Duchene, D.; Corstensen J.T. Drug Dev. Ind. Pharm. 1979, 5(3), 227.
35. Nakabayaspi, K.; Shimamoto, T.; Mimac, H. Chem. Pharm. Bull. 1980, 28, 1090-99.
36. Purwadaria, H.K.; Heldman, D.R.; Kirk, J.R. J. Food Proc. Eng. 1979, 3, 7.
37. Tubert, A.H.; Iglesias, H.A. Lebensm Wiss. Technol., 1986, 19, 365-68.
38. Bizot, H. in COST 90 Final Seminar, Leruven, Belgium, 1981, Part WA 3.
39. Van den Berg, C. In Properties of Water in Foods; Simatos, D., Multon J.L. Eds.; Nijhoff Publishers: Dodrecht, Netherlands, 1985 p. 119-31.

40. Weisser, H. in Properties of Water in Foods; Simatos, D., Multon, J.L. Eds.; Nijhoff Publishers: Dodrecht, Netherlands, 1985, p. 95–118.
41. Lomauro, D.J.; Bakshi, A.S.; Labuza, T.P. Lebensm. Wiss. Technol. 1985, 18, 111-7.
42. Lomauro, D.J.; Bakshi, A.S.; Labuza, T.P. Lebensm. Wiss. Technol. 1985 18, 118-24.
43. Labuza, T.P.; Saltmarch, M. in Water Activity: Influences on Food Quality; Rockland, L., Stewart, G.F. Eds.; Academic Press: New York, 1981 p. 605-50.
44. Mizrahi, S.; Labuza, T.P.; Karel, M. J. Food Sci. 1970, 35, 799.
45. Mizrahi, S.; Karel, M. J. Food Sci. 1977, 42, 1575.
46. Mizrahi, S.; Karel, M. J. Food Proc. Preserv. 1977, 1, 225-234.
47. Mizrahi, S.; Karel, M. J. Food Sci. 1977, 42.
48. Cardoso G.; Labuza, T.P. J. Food Technol. 1983, 18, 587.
49. Peppas, N.; Klime, D.F. Proc. ACS Div. Polym. Mat. Sci. Eng., 1985, 52, p. 579-83.
50. Gencturk, M.B.; Bakshi, A.S.; Hong, Y.C.; Labuza, T.P. J. Food Proc. Eng. 1986, 8, 243-61.
51. Crank, J. The Mathematics of Diffusion, 2nd Ed.; Oxford University Press: London, 1975.
52. King, C.J. Food Technol., 1968, 22, 509.
53. Schwartzberg, H.G. J. Food Sci., 1975, 40, 211.
54. Bluestein, P.M.; Labuza, T.P. AIChE J., 1972, 18, 706-12.
55. Taoukis, P.S.; El Meskine, A.; Labuza, T.P. Inst. Food Tech. 46 Annual Meeting Abstr. 1986, p. 182.
56. El Meskine, A., M.S. Thesis, University of Minnesota, 1987.
57. Lomauro, C.J.; Bakshi, A.S.; Labuza, T.P., J. Food Sci., 1985, 50, 397.

RECEIVED September 24, 1987

Chapter 20

Emerging Technology
at the Packaging–Processing Interface

Aaron L. Brody

Schotland Business Research, Three Independence Way, Princeton,
NJ 08540

This review enumerates the various contemporary food
packaging technologies in which processing, product
and packaging are interdependent; describing the
technologies, the interactions and issues; and
providing references. Brief discussions are given
on some emerging future food packaging technologies
in which the interactions are anticipated to be
crucial.

In the early history of food technology, except for canning and
glass packing, packaging was an incidental operation. Food
technologists recognized, however, that the canning process was an
integration of package, product and process, and so the two were
developed as one from the early years of the twentieth century
until the 1960's. Subsequently, and until the early 1980's,
technical and economic rifts among industrial and research
interests led to sharp division between food processing and food
packaging. As a result of severe issues and potential problems
which surfaced as a result of the commercial introductions of
retort pouch, hot fill pouch, aseptic packaging, and modified
atmosphere packaging, a heightened awareness of a need to link
food processing, product and packaging has emerged. Among the
interactions of concern have been migration of plastic components
into foods from packaging materials; removal of desirable food
components by the packaging materials; heat seal integrity;
package integrity; chemical sterilant residuals; oxygen
transmission; and the creation of anaerobic conditions within
unsterile packages.
  Not all of these issues have been addressed by regulatory
authorities nor thoroughly investigated by independent research
groups. Since most of the problems became visible after
commercialization, and relatively few are regarded as potential
safety hazards, research to characterize and solve the problems
has been relatively sparse. The benefits potentially attributable
to full-scale commercialization is driving government agencies,

independent laboratories, universities and industry to study these
and other food processing: packaging relationships more closely.

The bulk of the valid reference material on these topics is
found in the presentations and proceedings of dedicated
conferences sponsored by for-profit organizations because the
technologies have been industrialized with relatively little
peer-review research. As more independent research is conducted,
the information base will be completed. This situation is not far
different from that of canning, when technologists from can making
companies and trade associations wrote the literature based on
their own research.

Shelf Life

Concepts of shelf life are changing as a result of distribution
economics, raw material sources, marketing needs and technology.
In Western economies, long shelf lives in excess of one year, and
very short shelf lives of less than two weeks are becoming
exceptions. Food industries are tending towards a shelf life mode
in the vicinity of six months.

Distribution Economics. Costs of inventory are now measurable and
are recognized as important. With computer-based control,
inventories of packaged foods may be minimized without adversely
affecting the ability of food processors and distributors to
deliver product in a timely fashion.

Raw Material Sources. In the past, much food was processed at or
near agricultural sources, with seasonal packing the accepted
practice. Increasingly, food raw materials such as tomatoes are
partially processed in bulk during harvest season, stored, and
withdrawn throughout the year for further processing.

Further, large quantities of food raw materials such as
tomato puree, concentrated juices and wine are .pa imported from
off-shore sources throughout the year and used as the base for
food products.

Thus, broadening of the geographic and time range for foods
and their ingredients has supported the business concept of
reducing inventories and distribution times.

Marketing Needs. Decreasing available retail shelf space, even
for basic foods, has drawn marketers to a variety of strategies to
maintain and extend retail market position. More products and
product forms with lower volumes and shorter runs have become
common practice. Consumer desires for higher quality; to try new
foods; and convenience of storage, use and consumption have led
food marketers to double the rate of new food product
introductions since 1980 and to far outstrip the capacity of
retailers to accommodate the product array available (1).

Technology. Both food and packaging technology have
contributed to the altered view of shelf life. Marketing needs
appear to be dividing food quality into several classifications:
premium quality; high quality; good quality; standard; and

sub-standard, i.e., safe, edible and nutritious. Retention of
quality is based on the target market and consequently the
technologies selected to meet the needs.

The extreme perishability of so-called "fresh" foods has been
extended by raw material selection, rapid partial processing,
hygienic handling, packaging and controlled distribution (1).
Shelf life of dairy products including fluid milk is prolonged by
processing and packaging under aseptic-type conditions, but
maintaining low temperature distribution. Quality of both fresh
and processed meats is extended by modified atmosphere/vacuum
packaging coupled with low temperature distribution (2).

Both of these food categories, as well as others, are now
capable of shelf lives of up to six weeks as contrasted to one to
two days for fluid milk or a few hours for freshly ground beef
less than a generation ago. These technologies are representative
of the evolutionary processes that are applying the principles of
biochemistry, enzymology, microbiology and packaging to prolong
shelf life of highly perishable foods in directions that permit
regional and even national distribution.

Each technology integrates multiple scientific and
engineering principles to achieve the marketing and distribution
objectives.

Open Dating. Beneath the quantitative shelf life objective is the
now-accepted practice of open date coding. Despite the
well-documented difficulties of ascribing an accurate visible
package date code to a food product, most major food processors
and marketers are indicating pull and "best-by" dates. With just
a few days' shelf life, assurance that over-age inventories do not
exist depends on the shelf life extension technologies.

Shelf Stable Foods:

Shelf-stable is a term usually describing foods processed,
packaged and distributed to retain quality for periods of months
under ambient temperatures.   Included are canned, dry and
semi-moist foods. In this review, those shelf stable foods falling
into the retort pouch/tray and aseptic packaging categories are
specifically discussed in separate sections.

Canned Foods. Included in canned or thermally processed foods are
juices and juice drinks packaged in two piece aluminum cans with
liquid nitrogen to vaporize and generate an internal pressure to
maintain package shape and secondarily to help retain product
quality by reducing product oxidation. The initial reason was to
reduce packaging costs by employing cans manufactured in very high
quantities for carbonated beverages and beer.

By altering product pH and filling hot, reduced thermal
treatment pasta entrees in glass jars have been developed.

Confectionery. Although infrequently classed as a product
category benefitting from processing: packaging technology,
consumer acceptance of granola-type confections which are moisture

and oxygen sensitive has led to a reassessment of product shelf life for the entire category.

Because of their environmental sensitivities arising from their whole grain and fruit content, granola-type confections are packaged in heated sealed aluminum foil lamination. Aluminum foil offers maximum moisture barrier at low cost. When the foil is protected by plastic and/or paper, the barrier properties of the foil are maximized. Foil cracking particularly in corners permits minute openings through which moisture can enter. Interior heat sealing polyolefin coatings have been demonstrated to be capable of adding undesirable flavors and/or to scalp flavor notes from the contained product.

Mainstream confections such as coated chocolate bars undergo flavor staling as a result of exposure to heat, moisture and air. Glassine provides grease resistance and stiffness. With coating and heat sealing, glassine provides a modest moisture barrier. The introduction of cavitated core or pearlized oriented polypropylene has provided a base sheet with the stiffness, grease resistance and opacity of glassine plus inherent moisture barrier. When coated, flavor barrier is imparted, and a cold cohesive sealant may be applied. Although higher cost than equivalent glassine, coated coextruded opaque polypropylene has captured the majority of candy bar wrapping market by providing greater moisture and off-flavor protection and hence better shelf life. Further enhancing the barrier properties of the packaging materials are vacuum metallized polypropylene and polyester films (3).

Dry Mixes. Since their inception, dry bakery mixes have been protected from moisture. With the marketing objectives of quality niches, premium and high quality, dry mix technologists became aware of flavor nuances that establish quality differentiation.

By incorporating flavor barrier plastic such as nylon into coextruded polyolefin films, moisture proof pouches of dry mixes may have significantly superior flavor barrier and hence quality retention.

Cookies. The introduction of soft cookies required techniques to retain the textural and bite qualities of the original baked product. Unlike dry cookies which could tolerate modest moisture gain or loss, the quality of soft cookies is significantly altered by the addition or subtraction of moisture. Further, as a result of higher moisture content, fat oxidation rates are increased.

The commercial introduction of soft cookies which are distributed through retail grocery warehouse channels dictated high moisture/high oxygen barrier packaging coupled with inert atmosphere with the package. Further, since the cookies are fragile, physical protection must be integrated into the packaging.

Soft cookies are placed in compartmented thermoformed plastic trays which are, in turn, hermetically sealed in printed aluminum foil lamination bags. The bags are squared to help protect the cookies and improve shelf display. Flexible structures used include combinations of nylon or polyester film or paper on the exterior, aluminum foil and polyethylene or polypropylene on the interior. Nylon or polyester film ensures the gas barrier of the pinhole-prone aluminum foil when an inert atmosphere pack is used. Polypropylene on the interior eliminates plastic: cookie flavor interactions.

Thermally Stabilized Foods:

Detailed in this section are those foods with water activities above 0.85 which are heated to destroy microorganisms and enzymes and hermetically packaged before or after heating for long term ambient temperature shelf stability: aseptic hot filled and retort pouch/tray packaging. Not included here since the final result is not ambient temperature stability is thermal pasteurization for the purpose of extending refrigerated shelf life, a subject discussed in the section on modified/controlled atmosphere/vacuum packaging.

Aseptic Packaging. Defined as the independent sterilization of food and packaging and assembly under sterile conditions, aseptic packaging has ranged from ultra-clean in glass bottles and jars to superheated steam to sterilize cans in sizes ranging up to 10 gallons to sanitation of multi-thousand gallon tanks to the more widely known use of hydrogen peroxide to sterilize packaging materials prior to delivery of sterile liquid product. Sterilization of products prior to introduction into the aseptic packaging systems has been generally by continuous heat exchange systems to achieve high temperature/short time (HTST) or ultra high temperature/ultra short time (UHT) thermal conditions. The thesis is that since microbicidal and sporicidal effects take place at logarithmic rates and biochemical effects at geometric rates, the faster the heating, the greater the desired sterilizing effect while minimizing the cooking or biochemical effect (4-5).

For milk and dairy-based products, which are heat sensitive, and for most juices and juice drinks, the theory was proven. With all products, and especially fluid milk, however, the biochemical changes subsequent to packaging were unlike those experienced before in short-term storage refrigerated distribution or in long-term storage after excess thermal treatment. In some instances, the changes can be attributed to insufficient enzyme destruction or enzyme regeneration, an effect that accelerates oxidative browning and off- flavor developments. In other situations, the presence of minute quantities of residual dissolved or occluded oxygen in the sterile product, while of little consequence in short-term refrigerated conditions or in long-term storage of already severely heated product, is a very significant deteriorative vector in long term storage for product that is supposed to retain initial high quality. Removal of occluded oxygen is recommended for products prior to aseptic packaging.

Headspace oxygen is another contributor to product deterioration. While some aseptic packaging systems seal the package through the product and thus preclude headspace oxygen, most other systems employ flowing inert gas which does not completely eliminate headspace oxygen. Pressurized gas flow does not wholly remove headspace air. Only those packaging systems initially engineered to omit headspace oxygen, e.g., filling and closing under inert atmospheres, effectively overcome the headspace oxygen problem. Non-oxidative biochemical reactions not identified in refrigerated or canned product storage are also very significant degradation vectors in aseptic products engineered for long term retention of initial good quality. These reactions are attributed to the proximity presence of larger quantities of reactants over extended time.

Although suspected for many years, only recently has the scalping of flavor from fruit-based products been demonstrated quantitatively as a major contributor to product quality alterations in storage. Polyolefin interior linings have been proven to remove desirable flavor attributes from juices and juice drinks (6–9).

Although the permeability of plastics to oxygen has been studied extensively, the major external oxygen source may be breaches of packaging integrity and consequent transmission rather than permeation. Seal integrity is a crucial issue relative to microbiological recontamination, and heat seals through thermoplastics have not proven as reliable as mechanical seals, a subject discussed below. When seals are not microbiologically sound, they do not exclude oxygen (10).

Further, where scored and bent, aluminum foil often displays cracking and consequent entry paths for oxygen (11).

Heat sealants such as polyethylene, polypropylene, ionomer or ethylene/vinyl acetate copolymer have high oxygen permeability. Although the area facing the air is very small and the distance to travel is very long in heat seals, oxygen has a path and is able to enter heat sealed packages. Passage edgewise through thermoplastic heat seals has been proven to be a significant source of oxygen to contained product.

In general, the structural strength of plastic and paperboard/aluminum foil/plastic lamination is less than that of metal cans and glass jars, and thus their ability to withstand impact vibration stress, compression, etc., is not great. Structural integrity may be impaired by physical abuse, and oxygen, as well as micro-organisms, may enter.

Thus, in aseptic packaging, product quality may be compromised by:

- failure to inactivate enzymes
- enzyme regeneration
- failure to remove oxygen from the product
- oxygen in the headspace
- oxygen permeation through the packaging material
- oxygen transmission through breeches in the packaging
- structural failure
- oxygen permeation edgewise through heat seals
- accelerated biochemical action due to high reactant levels

- inability of high temperature/short time thermal processes to deliver product with perceptibly better quality than by low temperature/long time thermal methods (12).
- residual hydrogen peroxide sterilant when this sterilant is used (13).

Low Acid Fluid Foods. Worldwide, fluid milk and analogues, almost all of which are low acid, represent the largest volume packaged under aseptic conditions. Despite the higher sterilizing values required, low acid fluid foods are readily sterilized in continuous heat exchangers.

Issues of seal integrity which may lead to microbiological recontamination are far more serious with low acid foods because of the potential for pathogenic growth and toxin or infection formation. Fillers that permit seal area product contamination; sealers that are not properly controlled with respect to temperature, pressure or time; undue stresses on the seals; inconsistent sealant distribution, etc., all can lead to compromises in heat seals and potential - microbiological contamination problems (10).

Low Acid Particulate Foods. Foods containing particles below 5 mm diameter behave thermodynamically like homogeneous fluids and so may be thermally sterilized in conventional continuous heat exchangers. As the minimum dimension increases, particles assume the thermodynamic characterization of solids and so fluid/solid heat transfer must be computed. Time/temperature considerations become increasingly difficult to compute when several different particulates, e.g., meat and potatoes, are involved. Heat transfer is facilitated by the presence of carrier fluids, but the heating and heat transfer characteristics of each of the components are different. Thermal inputs must be for the slowest heating component, which means that other components are necessarily overheated (13-16).

Because of the complexity of computing heat transfer into multiple particulates, in practice the entire mass is generally overheated to ensure sterility (17, 13).

Newly engineered continuous heat exchangers with larger entry and exit ports and wider rotator blade spacing are claimed to be able to permit heat transfer without undue physical damage to particulates passing through.

Assurance of the heat transfer into larger size, i.e., greater than 15 mm particulate foods, without physical damage appears to require discontinuous heat exchangers which are essentially batch heaters in which the product may be readily put in and discharged (18).

A further issue of sterilization of low acid particulates is that high temperature/short time heat treatment is not only not beneficial to many foods, it is actually detrimental. Meats such as beef have higher quality after low temperature/long time heating; shellfish often are not tenderized under high temperature/short time heating; many sauces require very long times to properly develop flavors (13, 19).

Aseptic Packaging: Future Considerations. Aseptic packaging has been demonstrated as advantageous for high acid fluids such as juices and juice drinks, provided the shelf life and particularly oxidation- sensitive shelf life is taken into consideration. With proper fillers capable of handling large particles, high acid particulate foods such as fruit can be benefited.

Employing extended thermal processes and higher pressure fillers, tomato-based products including those containing particulates, should become more significant commercially.

Notions of altering pH for the sake of accommodating for aseptic techniques do not recognize fundamentals that stipulate that process and packaging technologies should be directed to advantage and not the reverse. Aseptic packaging fits natural high acid fluid and fluidizable foods, when the oxygen is removed and kept out.

The ability to sterilize low acid particulates is questionable from a thermodynamic standpoint, and only slightly less marginal from a quality benefit perspective.

Aseptic packaging requires the perfect integration of a number of simultaneous and sequential unit operations. Failure of any can result in spoilage and economic problems for high acid foods and potential public health hazards with low acid foods (20).

The problems associated with aseptic packaging of low acid particulate foods coupled with the apparent marginal benefits do not signal a hopeful future.

Retort Pouch/Tray Packaging. Since the late 1940's, retort pouch packaging has been the subject of intense development with the objective of lower cost packaging and higher quality contents. The lower cost was to arise from the lower quantities of packaging materials, a fact that overlooked the complexity of the lamination required. The higher quality was to come from lower heat input from the higher package surface to volume ratio.

As was discussed above in the section on aseptic packaging, but preceding it historically, high temperature/short time heating does not necessarily deliver better initial quality food. Further, terminal heating after packaging accelerates interactions of food and interior sealants on packaging materials.

Except for military applications, retort pouch packaging has been a commercial failure for the following reasons:

- economics of materials over cylindrical cans have not been realized and are not projected to be achievable in the foreseeable future
- economics of filling and closing are significantly worse than for canning lines
- 100% inspection and quarantine are required for retort pouch packaging
- pressure override retorts with accurate controls are required
- structural integrity of retort pouches must be maintained by external secondary packaging
- quality benefits from high temperature/short time thermal processing are not universal

-   storage quality losses are significantly greater with retort pouches than with cans
-   food:packaging material interactions are greater with thermoplastic packaging materials than in cans
-   probabilities of microbiological recontamination and hence public health hazards are far greater with retort pouches than with cans
-   few, if any, perceptible consumer benefits

Early in the development of the retort pouch, the tray shape became interesting because of the ability to fusion seal around a relatively clean flange perimeter. Aluminum foil polypropylene lamination have been in tray forms have been commercialized with all of the limitations indicated for retort pouches. The two advantages are more reliable heat seal (10-11) and convenience of eating from the tray after cutting open the closure (18).

With the coincidental development of high oxygen barrier coextruded plastic sheet and domestic microwave ovens, a move began towards the integration of plastic trays and terminal sterilization. With only aluminum foil structures to act as heat seal closures, almost all of the problems of retort pouches are present with plastic retort trays plus:

-   slower heat transfer since plastic has relatively poor K value
-   limitations on the high temperature that can be used for heating due to softening and melting points of the plastic
-   difficulty of achieving a fusion seal
-   non-uniformity of distribution of the crucial high oxygen barrier plastic in the formed tray
-   non-uniformity of physical resistance of the tray due to fabrication limitations
-   necessarily shorter shelf life because of the plastic permeability
-   moisture sensitivity of the principal oxygen barrier plastic employed (21).

On the other hand, advantages were initially perceived:

-   the tray shape can be virtually any form to satisfy consumer and product needs
-   the convenience of eating from the tray is enhanced (22).

Neither of these proposed benefits was initially achieved. The impact of the microwave oven on the domestic kitchen generated another clear advantage for the plastic tray, a more logical means to reheat the retorted tray than immersion in boiling water. Further, retorted plastic trays do not suffer from the differential microwave energy absorption and consequent non-uniform reheating of frozen entrees and dinners. The imposed advantage of microwaveability brought an accompanying obstacle of having to fabricate plastic structures capable of dual ovenability, i.e, reheating in either microwave or conventional conduction/ convection ovens. Combination heat resistant/oxygen barrier plastic structures are difficult to fabricate (23).

Commercializations are being made with extraordinary caution due to the many problems that have surfaced, plus as yet incomplete resolution of problems such as heat sealing that had been anticipated.

The hopes for rapid acceptance by processors, packagers and marketers have not materialized and perhaps will not until effective and totally reliable technologies become available. With almost all the development now being conducted on a proprietary industrial basis, and with each organization uncovering and resolving each developmental issue on an individual basis, the route to widespread commercialization will be much longer than if the technological issues were aired openly so that all could benefit.

Quasi Retort Tray Packaging. Intruding into classification of retort tray is a group of low acid foods that are not retorted: pasta and filled pastas that are reduced moisture and are thermally treated at just below 100 degrees C. Packaged in plastic barrier materials using thermoform/vacuum seal techniques, the filled packages are exposed to sequences of microwave energy and hot air to develop uniform heat within the food mass and impart sufficient heat to destroy micro- organisms of concern. By maintaining the water activity below 0.85, the potential for growth of pathogenic micro-organisms without the vacuum package is reduced. The potential for spoilage is, however, present and microwave energy across an air boundary fosters non-uniform heating of the mass with potential stresses on the heat seal (24).

With no validated published reports on the microbiology of these food packages, no valid judgement can be made. This is another example of a technology which is being developed on a proprietary basis without the benefit of complete airing of key public safety issues.

Hot Fill. High acid fluids may be heated to sterilization temperatures within heat exchangers and then filled hot into packages. Exposure of the package interiors to the hot fluid contents sterilizes them, insofar as the entire package interior is exposed. Hot filling with or without some post-closure heating, followed by cooling, has been commercial for high acid products such as juices, juice drinks, sauces, etc., for many decades. In all instances, hot filling has been into metal cans or glass bottles/jars with inversion to ensure that shoulders, necks and closure interiors are subjected to thermal sterilization conditions.

Issues in hot filling include:

- assurance that the entire fluid volume has been exposed to sterilizing values
- assurance that the entire package interior has been exposed to the appropriate sterilization value before losing heat

-   removal of heat after sterilization to ensure that the
    product does not continue to cook or that heat resistant
    spores do not germinate
-   headspace oxygen
-   product overheating to achieve sterilization values

Application of hot filling to new and supposedly less expensive
plastic and plastic composite packaging materials and forms has
required addressing issues basic to all hot filling plus other
issues unique to the new packaging forms:

-   temperature limitation of the plastics in terms of softening
    and melting points
-   ability to invert or otherwise expose closure and headspace
    areas to fluid food thermal sterilization values
-   sensitivity of heat seal closures to hot fluids
-   headspace oxygen
-   interactions of hot fluids and plastic interiors
-   long term interactions of ambient temperature fluids to
    interior plastics
-   potential microbiological recontamination through compromised
    closures
-   oxygen permeation through plastic walls, especially when the
    barrier plastic materials are not uniformly distributed
-   hydraulic stresses on the heat seals during
    distribution vibration and impact

As a result of these issues, much food hot filled into plastic and
plastic composites is distributed under refrigeration. Among the
package forms now being hot filled are:

-   high barrier coextruded plastic bottles for catsup, barbecue
    sauce, jellies, relishes, etc.

    * polypropylene/EVOH
    * polycarbonate/EVOH/polypropylene

-   high barrier coextruded plastic cups with heat seal aluminum
    foil closure for apple sauce

    * polypropylene/EVOH
    * polystyrene/PVDC

-   aluminum foil lamination pouches in
    hotel/restaurant/institutional sizes for tomato- based
    products

-   gable top aluminum foil/polyolefin/paperboard lamination for
    juices and juice drinks

-   heat set polyester bottles for juices and juice drinks

-   PVC bottles for juices and juice drinks

The upper temperature limitations of PVC at 68 degrees C render this material somewhat questionable relative to assurance of sterilization value. The oxygen barrier limitations of the polyester and the PVC render the packages marginal from an oxygen permeation and therefore shelf life standpoint.

With proper attention to the problems, and especially sterilizing values in closure and headspace regions, and residual oxygen in the headspace, hot filling into plastic and plastic composite packaging can prove to be an effective commercial means to take advantage of the lower weight, cost and space volume of plastic packaging. In some instances, because it represents a relatively simpler technology than aseptic packaging, it can be introduced in place of or even possibly displace aseptic packaging.

Retortable Plastic Can. The problems cited above for heat sealing of plastic trays have led some cautious food packagers and their suppliers to develop high barrier plastic cans with mechanical or mechanically- assisted closures. Long experience with rigid metal end double seaming has demonstrated its reliability for seal integrity with metal bodies. Although the reliability with paperboard composite bodies is somewhat lower, the level is still better than with heat sealing in the context of retort trays.

Double seaming rigid metal ends to plastic bodies generally involves the fabrication of uniform body flanges followed by mechanical clinching of the end flanges to the body without damaging the plastic which often has a propensity to flow or crack under stress. Developments in Italy demonstrated that an injection molded cylindrical polypropylene can with tapered body flange could be successfully double seamed with a rigid steel end. Later work in England demonstrated that a thermoformed die-cut flange could provide an adequate basis for the application of an aluminum end. In more recent manifestations, coinjection blow molded wide mouth cans have been fabricated to accept aluminum ends with seam integrity reliability reported to be the same as that experienced with either steel or aluminum bodies, after careful double seam machine adjustments. Further, the seam integrity is reported as intact during and after retorting, attributable in some measure to engineering concentric collapsible rings into the base of the plastic can body (25-27).

By fabricating the can body from high oxygen barrier ethylene vinyl alcohol plastic plus desiccant plus polypropylene, a can with temperature resistance of up to 130 degrees C is possible. The desiccant in the structure aids in removing water from the moisture- sensitive plastic and helps preserve its oxygen barrier properties (28-29).

Shelf life of the contents in the barrier plastic can with seamed aluminum end is reported at 18 months as not different from that from a three-piece bimetallic can for contents such as lasagna, spaghetti and meat balls, chili con carne and three-bean salad. The product categories being canned are not noted for their sensitivities to deterioration. Further, in eight- ounce cylinders, the thermal sterilization processes do not induce high quality in the foods packed.

As with flexible lamination heat seal retort trays, these plastic cans are retorted under water with counterpressure to help maintain shape (28).

At present, the only visible advantage of the plastic can is its ability to be reheated in a microwave oven. The metal remaining after full panel pullout of the closure is too small a mass of metal to constitute a microwave radiation hazard. A secondary advantage might be convenience of eating directly from a can with a white interior and exterior, which would be the only benefit for bean-type salads which, of course, are not heated before consumption (28).

Bowl-shaped cans with double seamed rigid metal ends have been multi-layer melt phase thermoformed with apparently adequate flange uniformity to permit reliable high speed double seam closing. Also fabricated with EVOH as the sole oxygen barrier, these cans employ asymmetric placement, i.e., placing the EVOH layer away from the contents, to minimize the deteriorative effects of moisture on the barrier plastic (23,29).

Again, the food products contained do not suffer seriously from flavor deterioration in ambient storage.

Since both cans receive thermal sterilization values for solid pack low acid foods, the initial quality cannot be regarded as premium.

The major advantages of the bowl shaped plastic can are ability to be reheated by microwaves (abetted by a snap-on overcap that reduces the adverse effects of spattering) and convenience of eating directly from the package.

Development is underway in Sweden of a polypropylene can which employs aluminum foil in intimate contact with the plastic as the oxygen barrier. Closure is by an integration of mechanical clinch with induction fusion sealing of interior plastic liners.

It is instructive to review the sequence of driving forces for the development of the evolution of retortable pouches, trays and plastic cans:

Retort Package	Driving Forces in Sequence
- consumer size pouch	* economics * product quality
- military pouches	* volume * soft
- hotel/restaurant/ institutional pouches	* economics * volume * waste
- aluminum tray	* product quality * shelf display

– plastic barrier tray	* economics
	* uniqueness
	* ambient temperature shelf stability
	* microwaveability
	* consumption from package
– plastic cans, rigid metal closure	* economics
	* microwaveability
	* consumption from package

The sequence appears to emphasize a rationalization to employ aluminum foil lamination or plastic rather than a clear market need, i.e., supply side driven. The reasons for development of this evidently very difficult technology have not been obvious.

Carbonated Beverage/Beer Plastic Cans. The evolution of polyvinyl chloride to polyacrylonitrile to polyester for carbonated beverage bottles has been widely discussed. The further evolution of polyester from the original two-liter size to the present half liter for carbonated beverages despite the increased surface to volume ratio and to polyvinylidene chloride (PVDC) coated versions to contain English beer has also been well-documented (30).
Since the can shape involves a much smaller plastic surface-to-volume ratio, polyester was expected to have sufficient carbon dioxide barrier to contain carbonated beverages. The notion of containing pasteurized beer in any plastic has been repeatedly negated by brewery flavor and packaging experts. Polyester, even when PVDC coated, has insufficient oxygen barrier to avoid greater than one part per million oxygen in the can demanded to ensure against oxidative flavor change (31). Further, beer flavor interactions with plastic especially as a result of heat pasteurization are detectable, and, for the present, are not acceptable.
Polyester oxygen permeation, flavor contribution and flavor scalping are all satisfactory for most carbonated beverages including the dominant colas and fruit flavors. Further, polyester is capable of containing sufficient carbon dioxide for consumer acceptance, under tightly controlled distribution conditions. Polyester is also sufficiently heat resistant to resist the heat sanitation processes required prior to filling of natural juices which are being employed as flavor and mouth feel enhancers.
Polyester cans with aluminum double seamed ends have been demonstrated to be technologically feasible in commercial practice; carbonation retention, oxygen transmission and flavor interactions all proved acceptable. Polyester cans, whether produced by thermoforming or by injection flow molding, may be handled on moderate speed (i.e., 200 – 400 cans per minute) with satisfactory double seaming. The major question is the ability of suppliers to build capacity to meet the demands of high speed (i.e., 1600 cans per minute) canning lines. A subsidiary issue is the ability to recycle polyester cans with metal ends. This

corollary problem became the rationale for the original program
termination, but was not a deterrent to a renewed market test by a
tiny carbonated beverage canner.

The commercialization of polyester bottles and cans for
carbonated beverage containment implies a compromise in the flavor
integrity of the product derived from carbonation.

Modified/Controlled Atmosphere/Vacuum Packaging

Controlled atmosphere (CA) packaging, a term applied to the range
of controlled/modified atmosphere/vacuum packaging means that the
gaseous environment in immediate contact with the food is
controlled with respect to carbon dioxide, oxygen, nitrogen, water
vapor and trace gasses. Modified atmosphere (MA) is defined as an
initial alteration of the gas followed by time-based alteration
stemming from product respiration, microbiological action, package
gas transmission, etc. Vacuum packaging is removal of air from
the package with no gas replacement. All gaseous environment
changes may help extend shelf life by mass action biochemical
kinetic laws that lead to reduction in aerobic reaction rates with
reduction in oxygen, and, for CA and MA, by dissolution of carbon
dioxide in the water to alter pH and consequently slow reaction
rates. Further, as moisture content is elevated, evaporative
losses from the food are reduced and consequently the reaction
rates are simultaneously slowed. Respiration rates and
microbiological growth rates may be reduced by factors of two to
ten, with the greatest effects occurring at temperatures of 0 - 10
degrees C. Thus, the beneficial effects of CA/MA/vacuum packaging
for respiring foods or those susceptible to microbiological
deterioration are a result of complementing some other shelf life
extension mechanism. CA/MA/vacuum for such products is almost
ineffective alone, but in conjunction with refrigeration plus
other variables, is capable delivering shelf life extensions in
the range of 1.5 to ten times that in air at the same temperature
(32-33).

More than any other new food packaging/preservation
technology today, CA/MA/vacuum packaging must be an integral
component of total food systems, or else it can produce foods that
are potential public health hazards and foods whose quality is
actually deteriorated by the altered gaseous environment.

In the absence of oxygen, respiring fruit and vegetables
undergo anaerobic fermentations resulting in alcoholic, aldehyde
and ketone off-flavors that usually render the produce inedible
(34).

By suppressing the growth of aerobic micro-organisms and
permitting life of up to eight weeks, there is sufficient time
without evident spoilage to permit the growth of psychrophilic
pathogens such as species of Listeria and Yersinia whose growth
might not be slowed as much by CA/MA packaging. Since acid
formers, proteolytic organisms and visible mold grow under aerobic
conditions, in an air atmosphere, there is generally clear
evidence of spoilage well before pathogenic infections or
intoxications develop. Further, the expression of spoilage

microorganisms will usually dominate and prevent the growth of pathogens (35).

Under the anaerobic conditions ultimately created in some CA/MA/vacuum packaging in prolonged periods, aerobes are suppressed, permitting anaerobic spore forms and pathogenic species of anaerobes to grow and produce toxins. The CA/MA/vacuum conditions can be ideal for anaerobic growth and toxin production, particularly because long times are involved and since aerobes would not grow. It has been demonstrated that anaerobic pathogens could grow at temperatures as low as 4 degrees C and produce toxin without any sensory manifestation of food deterioration (36).

Minimizing problems with pathogens within CA/MA/vacuum packaged foods requires that the products be selected for highest microbiological quality, that they be processed and packaged under the strictest sanitation, that temperature reductions be rapid and that distribution temperatures be rigidly maintained as low as required to avoid anaerobic pathogenic growth. Compromises with these requirements could result in public health hazards in the packaged product (37-40).

Meats. Muscle food spoilage is generally microbiological. CA/MA/vacuum packaging retards microbiological growth, with vacuum more effective than CA/MA packaging.

Beef. Beef cut spoilage may be effectively retarded for up to six weeks at -1 to +3 degrees C by vacuum or MA packaging employing 20+% carbon dioxide in the environment. Aerobic microorganisms grow slowly and, although proteolytic enzymatic activity is retarded, it proceeds sufficiently to permit beef tenderizing. The sealed package retards water loss through evaporation. This system is employed commercially on a broad scale for distribution of primal and subprimal beef cuts from packaging house to retailer where final retail cuts are made. As is the rule with all CA/MA/vacuum packaging, optimum benefits are achievable only when strict hygiene and temperature controls are observed (2).

Under reduced oxygen conditions, the muscle pigment myoglobin is purple color. Only when exposed to oxygen is the typical cherry-red oxymyoglobin formed. Thus, vacuum packaged beef is purple. To obtain the red color claimed to be desired by consumers, the beef is packaged under 20-40% oxygen / 20-40% carbon dioxide/balance nitrogen, a combination which fosters maintenance of red color while simultaneously suppressing microbiological growth, provided all other sanitation and temperature controls are enforced.

The growth of microorganisms in ground beef is accelerated by the grinding process which distributes the cells throughout the mass. Shelf life extensions is achievable by MA, low oxygen or vacuum packaging, with only MA permitting the desired red color. Upon exposure to air, color of low oxygen or vacuum packaged ground beef reverts to oxymyoglobin red. Effective shelf life of ten or more days is achievable by using beef cuts and trimmings with low initial microbiological counts, by grinding cold, and by rigid adherence to temperatures below +4 degrees C throughout distribution. In commerce, beef may be coarse ground and packed under low oxygen for shipment to retailers where final grinding takes place; may be fine ground and packed under low oxygen conditions in opaque high oxygen barrier films for retail sale; or may be ground and packaged under high oxygen MA for retail distribution.

Despite the marketing admonition against the purple color of beef at retail level, several high oxygen barrier skin and vacuum packaging systems for retail beef cuts have been commercialized in the United States. The microbiological issues are magnified because the packages are out of packer and retailer control upon leaving the retail establishment.

Pork. With less of a color problem than beef, virtually no ground pork product and a more oxidation- prone fat, pork is more easily packaged under reduced oxygen. On the other hand, pork appears to be more prone to discoloration from excess carbon dioxide and to microbiological deterioration. Therefore, sanitation and temperature control, particularly in removing body heat from the carcass as soon as possible after slaughter, are even more important than with beef.

Commercial pork packaging involving CA/MA/vacuum appears to be tending more in the direction of vacuum packaging of layer cuts and of conventional packaging of multiple retail cuts in bulk MA bags or bag-in-box. By employing this method, consumers see familiar packaging on retail display, but the shelf life extensions of reduced oxygen are realized.

Poultry. The microbiology of poultry includes the almost ubiquitous presence of Salmonella species. Thus, poultry packaging mandates a rapid removal of body heat and maintenance of temperatures of -1 degrees C to +1 degrees C. At these temperatures, sufficient shelf life for commercial distribution turnovers of under two weeks are achieved. Cut-up chilled poultry is packaged for conventional retail display and multiple bulk packed for distribution to retail stores. The incorporation of a carbon dioxide/nitrogen atmosphere within an oxygen/carbon dioxide permeable package handled under the same temperature conditions as air- packed adds another two to four days shelf life. Gas permeable film bags permit same loss of carbon dioxide and ingress of oxygen to avoid anaerobic microbial growth (41-44).

Fish. Much edible fish in the United States comes from the ocean bottom where the fish can become contaminated with Clostridium botulinum Type E. This pathogen is capable of growth and toxin formation at temperatures as low as 5 degrees C with no other signal of spoilage. Because of the clear danger of toxin formation under normal U.S. temperature distribution conditions which range above 10 degrees C, FDA and Department of Commerce have, with a few exceptions, discouraged MA/CA/vacuum packaging of fish. The exceptions involve vacuum packaging in sealed oxygen permeable materials of selected microbiologically clean fish distributed under tight low temperature control (45,38). In

Western Europe, these restrictions are far less rigid and MA packaging of fish and shellfish in not uncommon.

Bakery Goods. Soft baked goods such as bread and cakes undergo staling on textural losses as a result of starch retrogradation. Staling increases with decreasing temperature and is in perception retarded by elevated moisture, a variable than encourages the growth of surface molds. Carbon dioxide concentrations above 20% effectively retard the growth of surface mold by factors of 1.5 to 2 and thus permit sealed packaging with high internal relative humidity. The presence of excess mold cannot be overcome by carbon dioxide, however, and so the bakery product must be packaged under clean conditions immediately following baking with minimum or no exposure to the ambient air (46).

Because of the large internal void volume, excess carbon dioxide must be introduced, even to the extent of injecting the gas into the crumb (46–47).

By employing thermoform/vacuum seal packaging equipment, bakery aromas are lost in the vacuum withdrawal phase prior to back flushing with gas (48).

An interesting variation that compensates for most of the problems is partial baking in the package. Carbon dioxide and water vapor generated during baking expel oxygen and create the internal controlled atmosphere. The package is sealed immediately and condensing internal gases draw the packaging material tightly to the bread or rolls. If the plastic does not interact with the product, and if the seals remain intact during the dual stresses of internal pressure and vacuum, the product can experience a shelf life of up to three months. Because it is par-baked, the product is reconstituted in an oven by the consumer to deliver a bread or roll with a crusty exterior and a soft interior resulting from a brief moisture-driven rejuvenation from staling (46).

Some rye breads in West Germany are being packaged under CA to avoid the presence of chemical preservatives. Some pumpernickel breads are being long time- low temperature heated and packaged immediately to take advantage of multiple effects of reduced oxygen, elevated carbon dioxide and moisture and trace quantities of ethanol and acetic and formic acids, all of which are mold inhibitors (48).

Precooked Foods. By carefully integrating sanitation, low heat pasteurization and controlled atmosphere packaging, a new class of long shelf life refrigerated entrees, salads and side dishes is being developed (49). Because so many components including seasonings and spices are mixed in these products, they are contaminated with a wide variety of microorganisms whose count is only reduced, not eliminated by heat pasteurization. The products usually enjoy the absence of enzymatic activity but suffer from viable mixed microbial flora in good growth environments. With no atmospheric control, shelf life even under the best refrigeration, is on the order of three to six days. With controlled atmosphere, equivalent shelf life can be extended to three weeks or possibly longer (50).

Precooked low acid foods under controlled atmospheres constitute the product category having the greatest potential public health hazard. On the other hand, with proper selection of raw materials and processing, plus temperature control, product quality reaching the retail consumer can equal that of freshly prepared (51).

An uncompromising microbiological regimen must be present for ingredients, components, processing and packaging. Temperature control throughout and especially in reduction from pasteurization and in distribution must be strictly controlled.

The atmospheric environment within the package while important is less critical than sanitation and temperature control.

Among the systems in commercial development are:

-   "souse-vide:" cooking of individual components to a degree that accounts for later pasteurization and reheating; vacuum packaging in a barrier plastic tray (52)
-   cook-chill: blending and heating of components in a vacuum sealed barrier plastic or aluminum foil lamination pouch; chilling by ice water immersion or evaporative cooling (49)
-   chill and CA package
-   cook-chill: hot fill into pouches and chill in ice water or by evaporative cooling.
-   package in plastic trays and heat under steam or steam water (24)
-   package in barrier plastic trays and apply sequence of microwave and hot air (24)
-   Multitherm: package in barrier plastic trays and apply integrated microwave and conduction heat under pressure

Regardless of the implementation, the operational sequence is the same:

- assemble components
- package under reduced oxygen
- heat to pasteurization
- chill rapidly
- distribute under refrigeration
- maintain strict microbiological standards

Conclusion

Those food packaging technologies that are being developed to take advantage of new plastic and plastic composite materials are generally driven by supply-side objectives. Often, the projected economics and quality benefits have not been realized. Food product contents have been sometimes overlooked or compromised in the attempts to create packaging adapted for the packaging materials.

Among the issues that have not necessarily been always fully comprehended are:

- selection of food product components to accommodate the process or package
- microbiological quality of the food components and processes
- sterilization values
- food – plastic interactions, especially under elevated temperatures or in long-term distribution
- temperature control
- final product quality
- package integrity
- stresses on packaging during processing and distribution

Crucial to the resolution of issues is the integration of packaging and food technologies.

The slow progress being made in many of the new technologies reflects an increasing awareness that the food, microbiology and quality elements have only recently been recognized as significant and addressed. The issues emerging are so crucial to public health safety and to commercial success that individual organizations can no longer afford to maintain information proprietary. By openly sharing in problems and their resolution the new food packaging technologies can be efficiently developed.

Literature Cited

1. LeMaire, W. Proc. NOVA-PACK '87, Princeton, N.J.: Schotland Business Research, Inc., 1987.
2. Noyes, P. J. Proc. CAP '86, Princeton, N.J.: Schotland Business Research, Inc., 1986.
3. Kemp, B. C. Proc. TAPPI 1983 Paper Synthetics Conference, Atlanta, Ga.: TAPPI, 1983.
4. Carracher, D., Ed. Proc. Symposium, Campden Food Preservation Assn., 1983.
5. Anon. Capitalizing on Aseptic II--NFPA Conference, Washington, D.C.: The Food Processors Institute, 1985.
6. Abbe, S.; Bassett, B. M.; Collier, J. C. Proc. 1985 TAPPI Polymers, Lamination & Coatings Conference, Atlanta, Ga.: TAPPI, 1985.
7. Delassus, P. T. Proc. 1985 TAPPI Polymers, Lamination and Coatings Conference, Atlanta, Ga.: TAPPI, 1985.
8. Mannheim, C. H. Proc. ASEPTIPAK '85, Princeton, N.J.: Schotland Business Research, Inc., 1985.
9. Marshall, M. R.; Adams, J. P.; Williams, J. W. Proc. ASEPTIPAK '85, Princeton, N.J.: Schotland Business Research, Inc., 1985.
10. Downes, T. W. Proc. Packaging Alternatives for Food Processors, Washington, D.C.: National Food Processors Association, 1983.
11. Downes, T. W.; Goff, J.; Twede, D.; Arndt, G. Proc. ASEPTIPAK '85, Princeton, N.J.: Schotland Business Research, 1985.
12. LaGrenade, C.; Schlimme, D. V. Effects of Low Temperature Storage on the Shelf Life and Quality of Aseptically Packaged Foods, Research Paper 86-1, The Refrigeration Research Foundation, Bethesda, Md. 20814, July, 1986.

13. Madsen, A., Ed., Proc. ASEPTIC PROCESSING AND PACKAGING OF FOODS, Lund, Sweden, Swedish Food Institute, 1985.
14. Atherton, D. Aseptic Processing: A Study to Establish the Capabilities and Limitations of Available Machinery for Aseptic Processing and Packaging of Foodstuffs, Tech. Memo. No. 270, Campden Food Preservation Research Assn., 1981.
15. Holdsworth, S. D.; Nang, N. S.; Newman, M. Food Particle Sterilization, Tech. Memo. No 189, Campden Food Preservation Research Assn., Chipping Campden, U.K., 1978.
16. Knapp, M. E.; D. D. Disch. Proc. FUTURE PAK '85, Whippany, N.J.: Ryder Associates, 1985.
17. Adams, J. P.; Peterson, A.; Otwell, W. S. Food Tech., 36:4, 1983.
18. Wernimont, D. V., Proc. ASEPTIPAK '85, Princeton, N.J.: Schotland Business Research, Inc., 1985.
19. Ongley, M. H.; Adams, J. B. HTST Processing/Aseptic Filling of Particulate- Enzyme Problems, Tech. Memo. 297, Campden Food Preservation Research Assn., England, 1982.
20. Tarr, J. Proc. FOODPLAS '85/86, Hoboken, N.J.: Plastics Institute of America, 1986.
21. Bresnehan, W. Proc. FUTURE PAK, Whippany, N.J.: Ryder Associates, 1984.
22. Johns, D. Proc. FUTURE PAK, Whippany, N.J.: Ryder Associates, 1985.
23. Miller, R. W. Proc. FOODPLAS 85/86, Hoboken, N.J.: Plastics Institute of America, 1986.
24. Merrill, R. E. Proc. CAP '86, Princeton, N.J.: Schotland Business Research, Inc., 1986.
25. Wachtel, J. A. Proc. FUTURE PAK, Whippany, N.J.: Ryder Associates, 1984.
26. Gruber, W. Proc. FOODPLAS '87, Hoboken, N.J.: Plastics Institute of America, 1987.
27. Houtzer, R. Proc. FOODPLAS '87, Hoboken, N.J.: Plastics Institute of America, 1987.
28. Wachtel, J. A. Proc. PACK ALIMENTAIRE '87, Schotland Business Research, Inc., Chicago, 1987.
29. Salvage, B. Prepared Foods, 156, 7 July 1987, p. 152.
30. Brody, A. L. Brewers Digest, April, 1986.
31. Hardwick, W. A. Proc. COEX AMERICA '86, Princeton, N.J.: Schotland Business Research, Inc., 1986.
32. Bell, L. Proc. PACK ALIMENTAIRE '87, Schotland Business Research, Inc., Chicago, IL.
33. Saunders, R. Proc. PACK ALIMENTAIRE '87, Schotland Business Research, Inc., Chicago, 1987.
34. DeLeiris, J. Proc. NOVA-PACK '87, Princeton, N.J.: Schotland Business Research, Inc., 1987.
35. Mitchell, M. "An Investigation into the Microbiological Status of Commercially Prepared Salads," Atlanta, Ga.: Salad Manufacturers Assn., 1987.
36. Post, L. S.; Lee. D. A.; Solberg, M.; Furgang, D.; Specchio, J.; Graham, C. J. Food. Sci., 50(4):990-996, 1985.
37. Brody, A. L. J. Pkg. Tech., 1, 2, 2:39, 1987.
38. Fain, A. Proc. CAP '86, Princeton, N.J.: Schotland Business Research, Inc. 1986.

39. Silliker, J.; Wolfe, S. Food Tech., 34, 1980.
40. Hintlian, C.; Hotchkiss, J. H. Food Tech., 40, 70, 1986.
41. Baker, R. C.; Qureshi, R. A.; Hotchkiss, J. H. Poultry Science, 64:328-332, 1985.
42. Hotchkiss, J. H. Proc. CAP '86, Princeton, N.J.: Schotland Business Research, Inc., 1986.
43. Goss, R. R. Proc. CAP '86, Princeton, N.J.: Schotland Business Research, Inc., 1986.
44. Goss, R. R. Proc. PACK ALIMENTAIRE '87, Schotland Business Research, Inc., Chicago, 1987.
45. Regenstein, J. M. Proc. CAP '86, Princeton, N.J.: Schotland Business Research, Inc., 1986.
46. Seiler, D. A. L. Proc. CAP '84, Princeton, N.J.: Schotland Business Research, Inc., 1984.
47. Benson, A. Proc. CAP '86, Princeton, N.J.: Schotland Business Research, Inc., 1986.
48. Brummer, J. Proc. CAP '86, Princeton, N.J.: Schotalnd Business Research, Inc., 1986.
49. Otto, E. Proc. PACK ALIMENTAIRE '87, Schotland Business Research, Inc., Chicago, 1987.
50. Louis, P. J. Proc. CAP '84, Princeton, N.J.: Schotland Business Research, Inc., 1984.
51. Girardon, P. Proc. CAP '86, Princeton, N.J.: Schotland Business Research, Inc., 1986.
52. Pralus, G. La Cuisine Souse Vide, Briennon, France, Georges Pralus, 1986.

RECEIVED December 29, 1987

Chapter 21

New Packaging for Processed Foods

Opportunities and Challenges

Michael E. Kashtock

National Food Processors Association, 1401 New York Avenue, NW, Washington, DC 20005

Packaging in a Changing Food Industry

One of the most noticeable evidences of change in the processed food industry today involves the new forms of packaging that have begun to appear on store shelves. The universe of these new structures is characterized by terms such as retortable, dual ovenable, microwavable, aseptic, hi-barrier, co-extruded, etc. (1). To the consumer they offer convenience and quality attributes in step with modern lifestyle expectations and demographic factors. Consequently, the food processor must meet the challenges of these new technologies to prosper, and whoever does this well, may determine who remains viable in today's increasingly competitive marketplace.

Not so long ago packaging choices for processed foods were relatively few. Retorted and hot filled products were available in cans and glass containers of varying sizes. Frozen items were available in boil-in-bag type pouches and foil trays. These packages were proven performers and played an important role in building and maintaining the confidence of the American public in the safety and quality of processed food products.

However, the former state of affairs has given way to an array of new packaging choices for processed food products, and to understand why this is happening, one must consider the driving forces behind these changes. Economic factors have led the industry to pursue the use of lighter weight and unbreakable packaging materials in the interest of reducing shipping costs and liability risks. Perhaps the most visible success story in this regard has been the conversion of larger size carbonated beverage bottles from glass to polyethylene terephthalate (PET) over the last decade. This trend is continuing (with the additional driving force of user convenience) with the development of squeezable plastic barrier bottles for condiment items such as catsup, jelly, mayonnaise, relishes, toppings and sauces. The extremely successful 1983 market introduction of Heinz' catsup in a squeezable blow molded barrier bottle led the movement in this direction.

0097–6156/88/0365–0284$06.00/0
© 1988 American Chemical Society

It is generally agreed by industry marketing experts that demographic trends have resulted in an increases in the numbers of dual wage earner households, households headed by one wage earning parent, and households without children (either comprised of young career types or senior citizens). These developments have created an increasing demand for conveniently prepared food products, because in many such households time is at a premium.

One alternative for the time pressured meal server is the traditional stop at the fast food restaurant on the way home from work. Supermarkets are also competing for the convenience business with the advent of the pre-prepared food section where items such as soups and entrees can be purchased in a microwavable container for quick and convenient reheating at home.

Food processors are now also entering this market with an array of frozen, refrigerated and shelf stable products designed for fast heating in the microwave oven (often in the retail container) and convenient serving and cleanup. For frozen dinner and entree products, this has led to a major shift away from the traditional foil tray to microwave compatible plastic based structures. Trays fabricated from crystallized polyethylene terephthalate (CPET), thermoset polyesters and polycarbonate based coextrusions have been very successful in filling this niche (2-4). These types of structures have the additional advantage of being able to be heated in the conventional oven and are thus termed "dual ovenable."

Shelf stable (retorted) items are also beginning to compete in this category. They offer extremely rapid heating in the microwave oven and surprisingly high quality, because their low profile tray type packaging reduces the amount of thermal processing required to achieve commercial sterility. Packaging of this type must have sufficient oxygen barrier for a one to two year shelf life at ambient temperature, must be compatible with the processing environment of a retort and must maintain seal integrity during processing and distribution; a tough challenge. Coextruded barrier containers utilizing ethylene vinyl alcohol or SARAN (copolymers of vinylidene chloride produced by Dow) as the barrier material are currently the leaders in this area (5), but newer barrier constructions are under development, with one, SELAR PA, an amorphous nylon, produced by Dupont, capable of being formed into monolayer containers.

The microwave oven is one factor in this convenience driven market for new packaging whose importance cannot be overstated. It is currently in use in about 55% of U.S. households, with this figure expected to rise to 80% by 1990.

Lifestyle changes are also playing an important role in the changing face of food packaging. The importance of high quality to contemporary consumers has led to the development of many so called "upscale" products ranging from premium frozen entrees and dinners to microwavable ice cream sundaes. Packaging for such products must promote the quality image the product seeks to project. For a frozen dinner for instance, this might dictate the use of an attractive "table ready" tray that has the appearance of dinnerware on the table (6). Another challenge to be met.

Extended shelf life refrigerated products such as pasta salads, entrees and full dinners are also now beginning to appear in this quality driven category. They often feature controlled or modified atmosphere packaging used in combination with other control factors (see below) to achieve their extended shelf life. For example, General Foods CULINOVA dinners currently in test market, are refrigerated controlled atmosphere packaged meals with a dated shelf life of three to four weeks (7). Such products as these offer minimal preparation time in the microwave oven and appeal particularly to the consumer's perception of refrigerated products as a high quality product category.

Aseptic packaging is a technology wherein the product and package are separately sterilized, and the product is then filled into the package and the package sealed in a sterile environment. The product is commercially sterile (meaning that any pathogenic or other spoilage microorganisms have been destroyed) and shelf stable (does not require refrigeration or freezing).

The emergence of aseptic packaging in the U.S. represents an example of many of the above cited driving forces operating together. Economic savings are realized in many cases by the use of lighter weight packaging materials such as polymer/foil/paper laminations or coextruded container constructions (8). Many aseptic packaging systems are based on form/fill/seal technologies that eliminate the need to ship preformed containers to the processor. The processor receives sheet or roll stock at his plant which ships more densely and at less cost. The lightweight nature of these materials also results in reduced shipping costs for the finished product.

Convenience in the aseptic product category is offered by products such as snack puddings and dips in EZ open thermoformed containers, and the familiar single serving juices in brik style containers with the punch through attached straw; Instant snacks and lunchbox items for busy parents to serve.

Higher product quality may result from the use of ultra high temperature (UHT) processes for aseptic products, because these usually result in less heat induced loss of product quality versus processes that heat the product in its container.

Aseptic packaging is now spreading from predominantly fruit juices into other product categories such as soups, gravies, baby food, tomato sauce, dairy drinks and yogurt (9).

Package Requirements

It is evident that packaging is playing an enhanced role in bringing products to the consumer that are in keeping with today's marketplace demands. However, beyond understanding the market driven changes affecting food packaging, one should appreciate that packaging plays a crucial role in delivering a safe and wholesome product to the consumer. The major packaging challenge facing food processors today is to assure that package related factors that affect the microbiological and chemical integrity of food products are understood and controlled in the production, distribution and handling of products whether they be frozen, refrigerated or shelf stable (aseptic or retorted).

Frozen products rely on freezing to retard microbial growth and chemical degradation. The package is not generally called upon to form a microbial or gas barrier around the product. Its principal function is to physically protect the product under the stressful conditions of frozen distribution and sale. Also, if the product is intended to be cooked in its retail container, the package must not chemically adulterate the product through the migration of toxic or off flavor components from the package during the cooking process. Of course, the latter applies for any food product type that is processed or cooked in its retail container.

Extended shelf life refrigerated products utilize refrigerated distribution, usually in combination with other factors (such as control of pH, water activity package gas composition, use of chemical preservatives and partial processing) to retard microbial growth and chemical degradation, and thus achieve an extended shelf life. Such products are likely to be marketed as premium products from which the consumer expects a high degree of quality.

A successful package for this type of product must be capable of maintaining the controlled or modified atmospheric composition selected for the product, particularly if such control is crucial to the microbiological safety of the product. Much attention and research is now being focused on psycrotropic pathogens, a category of microorganisms of concern in regard to this type of food product (10). As more is learned about the factors that control their growth in food media under refrigerated distribution conditions, packaging may play an increasingly important role in assuring the quality and safety of such products.

Shelf stable products are rendered commercially sterile by heating the product to achieve the destruction of pathogenic microorganisms either: 1) within a sealed container, or 2) separate from the container followed by aseptic packaging of the product. Sterilization of the container may be accomplished by treatment of the container with heat, a chemical sterilant (e.g., hydrogen peroxide), ionizing radiation or combinations of these. The package for any shelf stable product must constitute a microbiological barrier to the exterior of the container. Maintenance of the product's commercial sterility depends upon this.

If the product's commercial sterility is compromised through a breach in the package's integrity, spoilage of public health significance could result depending on the type of microbial contamination that occurs, the nature of the product and the environment within the container. Because of the potential risks involved, the issue of package integrity for shelf stable food products is probably the most intensely scrutinized and cautiously approached challenge associated with new packaging technology in the food industry.

Challenges to be Met

This section will address in more detail some of the important package related factors that affect the microbiological and chemical integrity of packaged food products. From a food processors perspective these are factors that must be understood and controlled to assure the safety and wholesomeness of processed food products.

Seal Integrity. Seal integrity is critical to assuring the safety
of shelf stable products. Many new plastic container types utilize
heat to seal a lid to a container body (or to seal the container
itself in foldover fashion). Heat sealed containers may employ a
peelable or fusion (non-peelable) seal. The later, which will
generally result in a greater seal strength, can be utilized on
packages such as retortable pouches where the package is not opened
by peeling at the seal area. Peelable seals are being developed for
convenience oriented products, many of which are intended to be
heated and served in their retail container, such as microwavable
ready to serve soups. These seals are generally designed to be
opened by peeling them back from the container flange. For such an
application to be successful, this should be able to be done by the
average consumer as opposed to the household "Lurch" who is called
upon to open the pickle jar.
     A challenge that must be met is the development of lidding
systems employing a combination of materials and sealing parameters
(dwell time, seal pressure, seal temperature) that will result in
peelability and yet maintain seal integrity under the stresses
associated with retorting, distribution and handling. Packaging
material converters are developing such lidstock materials at the
present time. Several products are now in test market in the U.S.,
with larger scale commercialization anticipated in the near future.
     Standard peelable flexible lidding constructions are generally
laminated structures with a foil barrier layer, polymeric inner
(food contact) and exterior layers, and appropriate adhesives. A
new and somewhat different approach to lid construction recently
introduced by the Continental Can Company on its MENU BOX container,
utilizes a semi-rigid coextruded barrier sheet fabricated with an
integral pull tab. The lid is fusion sealed to the container flange
and undergoes separation at a material interface within the
coextruded sheet when the tab is pulled. Continental Can reports
that the tensile strength and seal strength of this structure
significantly exceed those of standard flexible lidstocks.
     No lidding/sealing concept is right for every packaging
application. Factors that should be considered in choosing a
packaging system include the type of product, its intended
processing, the type of secondary packaging to be used, and the
rigors of the distribution system. A thorough approach to assessing
a package's seal performance would include performing simulated
abuse testing and trial distribution shipments on commercial units
(i.e., palletized shipping units), followed by an examination of the
seal and the entire package (11).
     Processors should also have an ongoing quality control program
for seal evaluation of finished containers prior to commercial
shipment. Weak or incomplete seals can result from causes such as
sealing head pressure or temperature drops, inadequate sealing dwell
time, head misalignment, or product contamination on sealing
surfaces (12). At the present time many processors in test
marketing programs are relying on 100% visual inspection of
container seals, and some even hold the product for incubation
followed by a second examination before shipment. Line samples of
finished product should also be periodically pulled for seal
evaluation by a destructive technique such as burst testing (or dye

penetration or electroconductivity testing for aseptic brik type containers) (13). This documents the continuous performance of the sealing system within preset parameters.

Reliance upon manual visual inspection techniques will probably not suffice to handle commercial scale volumes of product. Automated systems are currently under development and evaluation that will provide for on-line high speed non-destructive seal testing. For example, one system, developed by Benthos Inc., operates by pressurizing a container's seal area and electronically monitoring for any deflection of the lid, which be indicative of a possible leak into the container from the seal area.

An industry committee (Flexible Package Integrity Committee) of the National Food Processors Association is scheduled to publish an integrity bulletin for certain types of flexible and semi-rigid packages in 1987. It will contain photographic illustrations of container defects with classification of the defects into the categories of critical, major, and minor, for heat sealed and double seamed plastic containers, retortable pouches and paperboard based containers used for aseptic packaging. This publication will be useful to personnel involved in container inspection and will serve as a first line sorting guideline for potentially defective containers prior to laboratory examination. It will also prove useful to personnel engaged in package development.

Another NFPA committee (Plastic Packaging Lab Methods Committee) is assessing the relationship between the manner in which destructive container testing is conducted and the results obtained. For instance, in burst testing, factors such as the rate of internal pressure increase in the container and the manner in which container expansion is restrained, influence test results. Data of this type generated for internal use or for submission to regulatory agencies should be based upon test procedures that minimize the effect of test variables.

Influence of the Retort Environment. The use of flexible and semi-rigid container types in retort applications requires an understanding of retort induced effects that may affect process adequacy and package integrity.

For example, containers build up internal pressure in a retort to an extent that depends upon factors such as the type of product, pre-processing treatments (blanching), the amount of entrapped gases in the product and the temperature within the container (14). If internal container pressure distorts the shape of the container during processing, the adequacy of the thermal process could be affected. In addition, excess internal container pressure could place undue stress on the heat seal. In retort systems, overpressure, provided by air, steam or steam/air mixtures is applied to prevent container deformation, and represents a critical process factor.

Because of the complex and critical nature of process related parameters, processors will generally call upon a recognized process authority to design and evaluate a thermal process in the product development stage. The process authority will also assist the processor in any required regulatory filing of the process.

     FDA regulations require that thermal processes for acidified
and low acid commercially sterile foods be filed with the agency
prior to packing the product, while the USDA's proposed canning
regulations require that processes be maintained on file for the
agency's inspection upon request. The United States Department of
Agriculture is responsible for the regulation of meat and poultry
products while the Food and Drug Administration is responsible for
all other acidified and low acid product types.

Aseptic Systems. In a commercial aseptic packaging operation for
low acid foods, it is of critical importance that the operator
control system parameters that affect package sterility and
integrity. Aseptic packaging equipment must be thoroughly tested
before commercial operation to assess the system's design and the
performance of the sterilization, filling and sealing operations.
The techniques for evaluating the equipment are quite sophisticated
and involve inoculating the packaging material and the equipment
itself with selected test organisms, followed by actual packing of
test product. The product is then microbiologically evaluated to
demonstrate that product sterility was achieved under operating
conditions (15).
     As with retorted products, government regulations require
process filing and the validation of process adequacy for aseptic
packaging equipment. Here also, processors generally seek the
assistance of a recognized process authority in performing system
evaluations.

Chemical Considerations for Packaging Materials. Under Section
201(s) of the Federal Food, Drug and Cosmetic Act (the Act) food
packaging materials are defined as food additives (when they may be
reasonably expected to become a component of the food), and must, in
accordance with Section 409 of the Act, be listed prior to use, in
the Code of Federal Regulations, in a food additive regulation
specifying conditions for safe use (Note: there are some exceptions
pertaining to generally recognized as safe (GRAS) substances and
prior sanctioned substances that are covered in Dr. Breder's
presentation in this symposium). Such a listing usually occurs in
response to the submission of a food additive petition establishing
the safety of the proposed use.
     Provisions in the food additive regulation may limit use of the
packaging material to contact only with certain types of foods, or
impose a maximum food contact temperature. The regulation may also
establish certain specifications for the packaging material.
Compliance with the applicable food additive regulation will largely
ensure the safety of a given use of a packaging material.
     However overriding the food additive provisions of the Act is
the so called general adulteration clause in Section 402 of the Act,
under which any potentially harmful substance in a packaging
material can render it adulterated, even if the material fully
complies with the applicable food additive regulation.
     For instance, residual impurities in substances used in the
manufacture of a polymer, if such residues are carcinogenic or cause
an off-flavor or odor in the food, could render the packaging
material adulterated. The food additive regulations state what the

chemical identity of the polymer must be, but usually do not stipulate how the polymer must be made, or what the degree of purity of the starting materials must be. Therefore, the method of manufacture of a resin as per the catalyst and solvents used, its curing and drying etc., has implications as to whether the resin is acceptable as a food grade packaging material, even if the resin fully complies with the applicable food additive regulation.

In this connection, some processors are electing to screen prospective packaging materials for volatile substances that may be cause for toxicity or off-flavor concerns using methods such as gas chromatography/mass spectrometry. Some are asking for more rigorous validation of letters of guarantee from their suppliers concerning the regulatory compliance of the supplier's packaging materials. Users and suppliers of packaging materials should have a mutual understanding as to what constitutes an acceptable packaging material for a given application.

Processors whose products will be processed or cooked in their retail container should also assure themselves that excessive migration of the container material does not occur at elevated food contact temperatures. Many of the existing food additive regulations for packaging materials predate the use of plastic containers as processing and cooking vessels, and do not contain upper use temperature limits, so the suitability of a container for any application involving thermal processing, hot filling, or home heating in container should be carefully evaluated. Such product/package interaction is likely to be product type dependent and may be assessed by conducting extraction studies with food simulating solvents or by sensory evaluation to ensure that the food itself does not taste like the package.

On the regulatory developments front, the FDA is currently developing a threshold of regulation policy that is intended to define conditions under which certain packaging materials may be used without requiring a formal food additive regulation (16). Such a situation would involve uses where the extent and nature of migration is such that it could be considered insignificant from a regulatory standpoint. This effort is of particular interest to packaging suppliers and users because it may ultimately eliminate the food additive petition requirement for certain package components such as new barrier materials used in coextruded structures, expediting their availability to the industry.

Sterilants Used for Aseptic Packaging. In aseptic packaging applications the treatment of the packaging material with chemical or physical sterilants (e.g., irradiation) also has implications concerning the food additive regulations. If a proposed sterilizing treatment could conceivably modify a regulated packaging material in such a manner that it no longer complies with the food additive regulation for the material, FDA could request data concerning such effect. If such treatment were shown to modify the packaging material, then a separate food additive petition covering the proposed use would likely be required. In the absence of any sterilant induced modification, the packaging material need only comply with the existing food additive regulation.

However, there is one important difference between the use of
chemical sterilants and irradiation concerning the food additive
regulations. Chemical sterilization of a packaging material is
likely to leave a residue on the packaging material which will
itself be subjected to clearance as a food additive (see the
regulation for hydrogen peroxide in 21 CFR 178.1005). On the other
hand, because irradiation of a packaging material separate from the
food is a physical process that does not leave a residue on the
packaging material, there is no requirement to submit a food
additive petition for the package sterilization process, provided
the process has not altered the packaging material.
     Irradiation of a food within its package is an entirely
different matter which would likely require a separate clearance
unless the packaging material has already been cleared for such use
under 21 CFR 179.45 (17).

Conclusion

Conquering the challenges of the new packaging technologies that
affect food quality and safety, will undoubtedly be major pursuits
of the food industry in the near and long term future. This paper
has largely emphasized the safety aspects of this subject, as this
is where many of the cooperative industry efforts coordinated by the
NFPA have been concentrated. We are pleased to see some of the fine
work directed at food quality and safety aspects being done at many
of our academic institutions, some of which was presented at this
symposium.
     In our opinion, some primary areas for ongoing research will
be: 1) ensuring the safety of food products that attain an extended
shelf life through the application of non-lethal processing and
packaging technologies such as modified atmosphere packaging, 2)
developing a better understanding of the factors that govern food
quality and shelf life and refining this understanding into better
techniques for modeling shelf life and accelerated shelf life
testing, 3) the evolution of packaging and products tailored
specifically for use with the microwave oven. Aspects such as
package design, construction and product formulation will be
engineered for optimum performance in the microwave environment, and
4) establishing the safety of many food packaging materials for
irradiation of food in the package. Such will be required for FDA
approval of new applications.
     Obviously, there is much that a user of the new packaging
technologies must understand. History has taught food processors
that they cannot simply be "users" of this new technology, but must
be informed and active users, continually seeking to learn more
about the relationship of the package to the product. This is why
the food processors of the NFPA have made a commitment to utilize
their resources in cooperation with suppliers and with each other,
to stay on top of the challenges that this new technology presents.

References

1. Packaging. January 1987, 32, 24.
2. Stras, J., Packaging, Encyclopedia. 1986, 31, 158.

3. Food Engineering. February 1986, 58, 38.
4. Food Processing. January 1987, 48, 160.
5. Packaging. December 1984, 29, 70.
6. Packaging. November 1985, 30, 27.
7. Packaging. April 1987, 32, 11.
8. Milgrom, J., "Cost Effectiveness of Various Plastic Packaging Materials", Foodplas 84/85, Plastics Institute of America, Inc.
9. Food Processing. June 1986, 47, 62.
10. Scott, V.N., "Control and Prevention of Microbiological Problems in New Generation Refrigerated Foods", Presented at 41st Annual Meeting of Research and Development Associates for Military Food and Packaging Systems, Inc., May 1985, Norfolk, VA.
11. DiGeronimo, M.J., "Dynamic Testing to Predict Package Performance in Distribution", Packaging Alternatives for Food Processors, National Food Processors Association.
12. Downes, T.W., "Q.C. Test Procedures for Seal Integrity", ibid.
13. Polvino, D.A., "Integrity Issues and Test Procedures for Semi-Rigid Retortable Containers", Presented at National Food Processors Association Annual Convention, February 1986, Atlanta, GA.
14. Polvino, D.A., "Retortable Plastic Containers Today: Technical Status", Proceedings of Foodplas IV/87, Plastics Institute of America, Inc.
15. Bernard, D., Gavin, A., Scott, V., Shafer, B., Stevenson, K., and Unverferth, J., "Evaluation of Aseptic Systems", Presented at Annual Meeting of the Institute of Food Technologists, June 1986, Dallas, TX.
16. Food Chemical News. February 10, 1986, 28, 71.
17. Federal Register. April 18, 1986, 51, 13394.

RECEIVED September 24, 1987

INDEXES

Author Index

Affiliation Index

Subject Index

Production by Barbara J. Libengood
Indexing by Colleen P. Stamm
Jacket design by Carla L. Clemens

Elements typeset by Hot Type Ltd., Washington, DC
Printed and bound by Maple Press, York, PA